ADVANCES IN CHEMISTRY RESEARCH

VOLUME 54

ADVANCES IN CHEMISTRY RESEARCH

Additional books and e-books in this series can be found on Nova's website under the Series tab.

ADVANCES IN CHEMISTRY RESEARCH

ADVANCES IN CHEMISTRY RESEARCH

VOLUME 54

JAMES C. TAYLOR
EDITOR

nova science publishers
New York

Library of Congress Cataloging-in-Publication Data

ISBN: 978-1-53615-828-1
ISSN: 1940-0950

Published by Nova Science Publishers, Inc. † New York

CONTENTS

PREFACE

The first chapter of *Advances in Chemistry Research. Volume 54* investigates the structure-function relationships of glycans. A recent structural analysis of oligomannose glycans, based on biophysical methods (including solution nuclear magnetic resonance and X-ray crystallographic analyses) are presented and discussed.

A summary of the history, clinical uses, and pharmacokinetics of polyethylene glycol interferon are presented. Important features regarding its clinical efficacy in combination with new antiviral agents like ribavirin, telaprevir, entecavir, tenofovir, adefovir, sofosbuvir and ledipasvir are also discussed.

The fluorescence properties of anthracene based ligands are proposed to be suitable to design sensors for cations as well as anions. Some of the complexes of anthracene-based ligands are reported to have potential for light-emitting devices.

Various quantum chemical properties, dipole moments, bond lengths, charges, total electronic energies, the highest occupied and lowest unoccupied molecular orbital energies and the interfrontier molecular orbital energy gap values are obtained and compared.

An empirical equation is derived for calculating the free formaldehyde concentration in alcohol–formaldehyde systems at various temperatures and total formaldehyde concentrations, as well as an equation for evaluating solvent concentrations.

In the final study, the authors exploit the intense fluorescence behavior of luminol for a quantitative understanding of hydrogen bonding behavior and its interaction with several biological media, with the help of steady state and time-resolved measurements as well as density functional theory calculations.

Chapter 1 - Glycans are diverse sugar polymers that play important roles in cellular and multicellular function. Monosaccharides are the building blocks of glycans and mannose is prevalent in and characteristic of mammalian asparagine-linked glycans (*N*-glycans). A trimannosyl core ($Man_3GlcNAc_2$-Asn) occurs in all types of *N*-glycan, including high mannose-type, hybrid-type and complex type. The mammalian high mannose-type *N*-glycan is composed of 5 - 9 mannose residues. The authors have been investigating the structure, dynamics and interaction of glycans, in order to better understand structure-function relationships. The authors' recent structural analysis of oligomannose glycans, based on biophysical methods, including solution nuclear magnetic resonance (NMR) and X-ray crystallographic analyses, are presented here and discussed with respect to structure and function.

Chapter 2 - Polyethylene glycol (PEG) is a versatile polymer not only used as an excipient in different dosage forms but it also plays an important role in the *in-vitro* and *in-vivo* stability of protein and polypeptide drugs. PEGylation helps to produce non-immunogenic, longer absorption and cost effective therapeutic proteins mainly indicated in cancer, hepatitis C, Crohn's disease, immunodeficiency disease, leukemia, and rheumatoid arthritis. Interferon (IFN) is a natural anti-viral agent. It stands second position after somatropin in worldwide prescription drug sales and investment and developmental strategies in the global biopharmaceutical market. In this chapter the history, clinical use, and pharmacokinetics profile of PEG-IFN has been presented. Important features regarding its clinical efficacy in combination with new antiviral agents, like ribavirin, telaprevir, entecavir, tenofovir, adefovir, sofosbuvir and ledipasvir especially in hepatitis is also discussed. The known and possible sites of PEGylation of

IFN are pointed out. The use of chromatographic techniques like, ion-exchange chromatography (IEC), hydrophobic interaction chromatography (HIC), size exclusion chromatography (SEC), and/or high performance liquid chromatography (HPLC) are mainly used for separation and purification of PEG-IFNs. Moreover, the novel site specific PEGylation with novel PEG derivatives are also described. The analytical techniques like reversed phase (RP) HPLC, gel electrophoresis, mass spectrometry are mainly used for quantitative and qualitative estimation of IFN and PEG-IFN. The stability profile of IFN and PEG-IFN is also discussed. Finally the chapter also through light on the future prospects on other PEGylated therapeutics proteins.

Chapter 3 - Anthracene has been functionalized with various donor groups such as azomethine, chalcogen, phosphorus, etc. to generate ligands of varying dentate character. The ligands are reported to react with metal precursors of different types, forming complexes of different geometries established by X-ray diffraction on their single crystals (if grown successfully). The fluorescence properties of anthracene-based ligands are suitable to design sensors for cations as well as anions. Schiff bases synthesized with anthracene skeleton have shown cytotoxic and fungicidal activity. Organometallic compounds designed with functionalized anthracene and other co-ligands are reported. MOF's are generated with anthracene-based ligands. Metal complexes of chalcogenated anthracenes efficiently catalyze *N*-alkylation of amines and transfer hydrogenation reactions. The catalytic activation of C-C coupling reactions with Pd complexes of anthracene based ligands is reported to be promising. Some of the complexes of anthracene-based ligands are reported to have potential for light emitting devices.

Chapter 4 - The germanium atom is substituted at three different positions (1, 2 or 9) of anthracene structure. The effect of centric perturbation arising from the replacement of carbon atom by a germanium atom at those positions of anthracene structure has been investigated within constrains of the density functional theory at the levels of B3LYP/6-31++G(d,p) and ωB97X-D/6-31++G(d,p). Various quantum chemical properties, dipole moments, bond lengths, charges, total electronic energies, the highest occupied and lowest unoccupied molecular orbital energies (HOMO and LUMO, respectively) and the interfrontier molecular orbital energy gap values, etc., have been obtained and compared. The germanium atom substitution perturbs the structure but planarity of the system is generally preserved. The isomeric germaanthracenes presently considered are structurally stable having reasonable bond lengths. The calculated infrared and ultraviolet-visible spectra of the perturbed anthracenes are presented. The NICS (0) values of the individual rings in each case have been calculated (B3LYP/6-31++G(d,p) level) and all these results are discussed.

Chapter 5 - A mechanism of the initiated nonbranched-chain process of forming 1,2-alkanediols and carbonyl compounds in alcohol–formaldehyde systems is suggested. The quasi-steady-state treatment is used to obtain kinetic equations that can describe the nonmonotonic (with a maximum) dependences of the formation rates of the products on the concentration of free unsolvated formaldehyde. The experimental concentration of the free unsolvated form of formaldehyde are given at the different temperatures, solvent permittivity, and total concentrations of formaldehyde in water and alcohols. An empirical equation for calculating the free formaldehyde concentration in alcohol–formaldehyde (including water/ethanediol–formaldehyde) systems at various temperatures and total formaldehyde concentrations and an equation for evaluating solvent concentrations in these systems were derived.

Chapter 6 - Luminol (5-amino-2,3-dihydro-1,4-phthalazinedione) is a versatile chemical that shows striking blue chemiluminescence in presence of certain metal ions when treated with appropriate oxidizing agent like hydrogen peroxide. This unique feature of luminol is often exploited by forensic investigators to detect trace amount of blood left in the crime scene. The chemiluminescence behavior of luminol has been extensively used by the biologists as a cellular assay to detect copper, iron and cyanides etc. It has also been applied to quantify the oxygen secretion by phagocytizing cells and detection of proteins, cancer biomarkers etc. An important aspect of luminol chemiluminescence is its varying degree of sensitivity from one substance to another. For example, it is very sensitive to animal or human blood, organic tissues and fluids in comparison with other compounds containing metal ions such as paints, metallic surfaces, household products, or vegetable enzymes. The proven criminological and clinical application of luminol demands more extensive research work on it in the areas of photochemistry, fluorescence spectroscopy and biophysical chemistry. A further point of interest is the possibility of using luminol as fluorescent probe, as it has been predicted to have hydrogen bonding characteristics akin to several biological entities. It may act as efficient fluorescent probe because of its sharp and strong absorption, as well as relatively high fluorescence quantum yield. Studying photophysical properties of model nitrogen heterocycles like luminol in homogeneous medium can lead to the understanding the comparative importance of solvatochromic parameters like α, β and π^* etc. on measurable spectroscopic quantities. The same processes investigated in the confined geometries ranging from "spheres" (micelle) and "cones" (cyclodextrin) to the more complex three-dimensional networks of proteins, DNA etc., would provide a unique opportunity that can lead to better understanding on the mechanism of drug recognition and action. In this contribution, the authors will exploit the intense fluorescence behavior of luminol for quantitative understanding of hydrogen bonding behavior and its interaction with several biological media with the help of steady state and time-resolved measurements as well as density functional theory calculations.

In: Advances in Chemistry Research
Editor: James C. Taylor
ISBN: 978-1-53615-828-1

Chapter 1

MANNOSE AS A COMPONENT OF GLYCANS: STRUCTURE, DYNAMICS AND INTERACTION

Yoshiki Yamaguchi[1,2,*] and Jun Uzawa[1]
[1]Structural Glycobiology Team, RIKEN, Wako, Saitama, Japan
[2]Laboratory of Pharmaceutical Physical Chemistry, Tohoku Medical and Pharmaceutical University, Sendai, Miyagi, Japan

ABSTRACT

Glycans are diverse sugar polymers that play important roles in cellular and multicellular function. Monosaccharides are the building blocks of glycans and mannose is prevalent in and characteristic of mammalian asparagine-linked glycans (*N*-glycans). A trimannosyl core ($Man_3GlcNAc_2$-Asn) occurs in all types of *N*-glycan, including high mannose-type, hybrid-type and complex type. The mammalian high mannose-type *N*-glycan is composed of 5 - 9 mannose residues. We have been investigating the structure, dynamics and interaction of glycans, in order to better understand structure-function relationships. Our recent structural analysis of oligomannose glycans, based on biophysical methods, including solution nuclear magnetic resonance (NMR) and X-ray

* Corresponding Author's E-mail: yyoshiki@tohoku-mpu.ac.jp.

crystallographic analyses, are presented here and discussed with respect to structure and function.

Keywords: *N*-glycan, high-mannose, NMR, X-ray crystallography, structure, dynamics, interaction

1. INTRODUCTION

Glycans are saccharide polymers playing major roles in important cellular events such as cell–cell interaction, cell signaling, and glycoprotein processing (Rademacher, Parekh and Dwek 1988, Varki 1993, 2017). Yet, information on structure-function relationships of glycans is rather limited, and we adopted a structural biology approach to gain a better understanding of glycan function. Glycans are typically glycoconjugates, covalently linked to proteins or lipids. Those attached to proteins are classified into two major types, *N*-linked (Asn-linked) and *O*-linked glycans. Mannose residues are typically found in *N*-linked glycans (*N*-glycans) and give rise to three types of *N*-glycans, namely high-mannose-, hybrid- and complex-types. All three have a trimannosyl core, with a common unit of ($Man_3GlcNAc_2$-Asn) and the structural variations of *N*-glycans are found in the outer part, beyond the trimannosyl core. High-mannose or oligomannose type structures of 5 - 9 mannose residues are formed initially in the biosynthetic pathway. They are converted to hybrid-type and finally to complex-type by glycan processing enzymes. Intracellular lectins interact with high mannose-type glycans to facilitate protein folding, transport and degradation. High-mannose type glycans often occur at specific sites on secreted proteins in the extracellular space or on the extracellular region of membrane proteins (Rudd and Dwek 1997, Suga, Nagae and Yamaguchi 2018). Structural and binding analyses of glycans at atomic level resolution are essential for detailed understanding of associated biological events and for the design of molecules that alter the system in desired ways. Over the past several years, we have been analyzing glycan structure, dynamics and interaction (Nagae and Yamaguchi 2015) and here present examples of our structural studies on oligomannose glycans using X-ray crystallography and solution NMR spectroscopy.

2. Basis of Oligosaccharide NMR

NMR is a unique technique in that atomic structural data can be obtained in the solution state. NMR signals need to be assigned before the detailed conformational analyses of oligosaccharides can proceed. Like other small organic molecules, ^{1}H signal assignments can be achieved by measuring two-dimensional NMR measurements, 2D ^{1}H-^{1}H DQF-COSY, HOHAHA(TOCSY) and NOESY(ROESY) via ^{1}H-^{1}H scholar coupling via chemical bonds and NOE(ROE) through space (explanation below). However, in practice, NMR signal assignment of oligosaccharides is challenging, even for small ones. A serious hurdle is the severe overlapping of proton signals. Typically the signal from anomeric protons (H1) is well separated from others, however, there is significant signal overlap among the rest of the protons (H2-H6). This hampers complete assignment of oligosaccharide NMR signals. The problem of ^{1}H signal overlap can be largely overcome by adding the ^{13}C dimension. ^{1}H-^{13}C HSQCs are generally used to discriminate anomeric signals from non-anomeric signals. Anomeric carbon typically resonates from 90 to 105 ppm while most non-anomeric carbons resonate at 60 - 80 ppm. The $^{3}J_{H1,H2}$ coupling constant is used to determine the anomericity of sugar residues. For example, an α-glucose residue typically shows a $^{3}J_{H1,H2}$ coupling value of 3 - 4 Hz, while β-glucose comes in at 7 - 10 Hz. In contrast, both α- and β-mannoses show small $^{3}J_{H1,H2}$ coupling constants of 1 - 3 Hz, and typically, α- mannose shows slightly larger $^{3}J_{H1,H2}$ values than that of β-mannose. The relatively small $^{3}J_{H1,H2}$ values of α- and β-mannose are due to a difference in the configuration at the 2-position between glucose and mannose. Hence the $^{3}J_{H1,H2}$ value cannot be used for identification of mannose anomericity. Alternatively, ^{1}H-coupled ^{1}H-^{13}C HSQC is often used to identify the anomeric configuration. The α configuration (equatorial H1) occurs at ~170 Hz as $^{1}J_{H1,C1}$, whereas the β configuration (axial H1) is at ~160 Hz.

^{1}H-^{13}C HMBC is a powerful method to identify sugar connectivites. HMBC cross peak interaction occurs between the anomeric proton (residue i) and the carbon atom of the connecting sugar residue (i-1), or between the anomeric carbon (residue i) and the proton of the connecting sugar residue (i-1). For example, in the 1 - 4 linkage, a cross peak will be observed between H1(i) and C4(i-1) or C1(i) and H4(i-1). For the 1 - 3 linkage, HMBC cross peak will be observed between H1(i) and C3(i-1) or C1(i) and H3(i-1). The intensity of the HMBC peak between H1(i) and C4(i-1) is dependent on the dihedral angle of H1(i)-C1(i)-O(i-1)-C4(i-1) in the 1-4 linkage. A large HMBC cross peak apprears when the torsion angle is 0° (eclipse) or 180° (anti) and is absent when it is 90°.

Interresidue NOE across the glycosidic bond is also important for correlating connecting sugar residues. In the case of a 1 - 4 linkage, interresidue NOE must be observed between H1(i) and H4(i-1) of the connecting residue. Additional weak NOE will be observed between H1(i) and H3(i-1) and/or H5(i-1), depending on the glycosidic torsion angles. These additional NOEs are helpful for defining the glycosidic conformation. It should be noted that strong interresidue NOE/ROE between H1(i) and H4(i-1) will be observed when the torson angle H1(i)-C1(i)-O(i-1)-C4(i-1) is 0° (eclipse). Therefore, the observation of an inter-residue HMBC peak in addition to an inter-residue NOE/ROE peak indicates an eclipse conformation.

The next step is to obtain the NMR observables which define the oligosaccharide conformation. Two important NMR parameters exist which provide quantitative structural information. One is the nuclear Overhauser effect (NOE). NOE is a through-space effect and is proportional to the inverse six power of the distance between the protons. Typically, if the distance is within 5 Å, an NOE will be observed. NOE is molecular weight dependent and under certain conditions, becomes zero. Typically, a NOE cannot be measured in a molecule with a molecular size of 800-1000 Da at an ^{1}H resonance frequency of 500-600 MHz. In such a case, ROE (rotating frame Overhauser effect) is used instead of NOE. A second parameter is spin-spin coupling, which is a through-bond effect. Importantly, the spin-spin coupling constant (*J*) is dependent on the dihedral angle between vicinal protons. Thus both distance and angular information can be obtained from NOE and *J*.

One serious limitation in the analysis of glycan structures is paucity of protons which define the oligosaccharide conformation. Typically only one interresidue NOE is observed for each glycosodic linkage. This is not enough to resolve the dihedral angle of each glycosidic linkage. In addition, because glycans are inherently flexible, a single conformation may not fully describe the structure. This can be addressed through other experimental and computational approaches.

3. Oligomannose Structure

High-mannose type glycans are presented on nascent proteins inside cells and are directly involved in quality control of the glycoprotein through interaction with intracellular lectins. Protein folding, transport and degradation events are facilitated by high-mannose type glycans and lectins, the glycan-binding proteins, recognize own-target glycans presented on proteins. Insight into how glycans accomplish these functions, can be achieved using structural biology approaches such as NMR and X-ray chrystallography.

Here we present solution NMR analyses of mono and oligomannose glycans as shown in Figure 1 (Ogawa, Katano and Matsui 1978, Ogawa et al. 1981, Winnik et al. 1982, Hori 1987, Hori et al. 1988, Kubota et al. 2012). First, signal assignment is necessary. A series of standard NMR measurements is applied, including 2D ^{1}H-^{1}H DQF-COSY, ^{1}H-^{13}C HSQC, ^{1}H-^{13}C HMBC and ^{1}H-^{13}C HSQC-TOCSY. One example is shown for the identification of mannose H2 protons in Man-4’, Man-D2 and Man-D3 residues in the pentasaccharide unit **8** (Kubota et al. 2012). ^{1}H-NMR spectra are shown, which were collected using double pulsed field gradient spin echo (DPFGSE)-double selective population transfer (SPT) difference spectroscopy (Figure 2) (Uzawa and Yoshida 2004). Double pulsed gradient spin echo (DPFGSE) sequence is one of the methods used to selectively excite a particular signal with the aid of pulsed field gradients. Two gradient echoes are used, rather than one, which gives more desirable phase properties than the simple gradient echo.

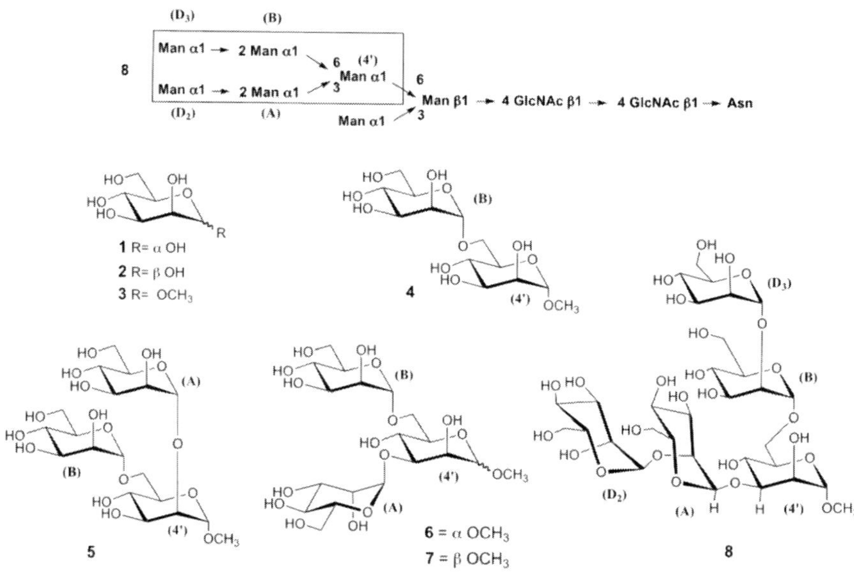

Figure 1. Structure of mannose monosaccharide and oligomannose glycans (**1**-**8**) described in this chapter. **8** is a part of the high-mannose type glycan and is shown within the rectangle.

This DPFGSE sequence can be turned into a selective inversion pulse (180° pulse) by simply adding a 90° pulse at the end. H1 protons of mannose residues are separated and can be selectively inverted. Selective inversion of each H1 signal clearly highlights each H2 signal by DPFGSE-SPT pulse sequence. The chemical shift of each H2 signal can be identified and the coupling consntants between H2 and H3 estimated as 3.0 Hz in each case. Typically the coupling constant between Man H1 and Man H2 ($^3J_{H1,H2}$) is very small (1 – 2 Hz). This SPT experiment was useful for connecting mannnose H1 and H2 even for such a small $^3J_{H1,H2}$.

Two protons, H6proR and H6proS are attached to C6 carbon of hexose residues such as mannose. Empirically, H6proR signal of an α1-6 linked mannose residue, such as Man-4', appear at lower field compared with the corresponding H6proS signal.

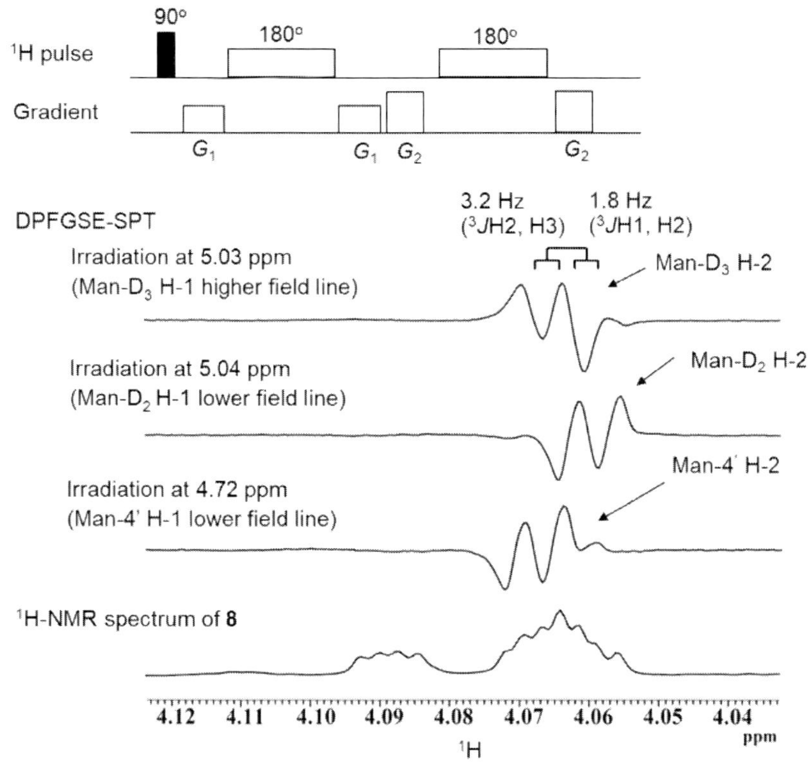

Figure 2. Selective excitation with DPFGSE unit. DPFGSE-SPT spectra of **8** are shown with different irradiation points. Control ^{1}H-NMR spectrum of **8** is shown below.

In constant, the H6proR signals of mannose reisidues without this 1 - 6 linkage, such as Man-A, Man-B, Man-D2 amd Man-D3, shifts to a higher field compared to the H6proS signals (Ohrui et al. 1985, Hori 1987, Nishida et al. 1988, Hori et al. 1988, Ohrui et al. 1991). ^{1}H and ^{13}C assignments of mannose and oligomannnose samples are shown in Table 1, and now the conformational analysis can be performed. Interresidue NOE/ROE and transglycosidic $^3J_{CH}$ are used for determining the conformatioanl restraints of the glycosidic linkages. Figure 3 summarizes the ^{1}H-^{1}H ROE and ^{1}H-^{13}C HMBC correlations observed in the oligomannose structures (**5-8**).

The oligosaccharide conformation will be further analyzed using molecular dynamics simulations in conjunction with the NMR data. Combined experimental and theoretical approaches are currently necessary to compensate for the limited number of experimental parameters attainable to completely define the oligosaccharide conformation and dynamics.

4. OLIGOMANNOSE DYNAMICS

To obtain information on dynamics, one simple method is to collect NMR spectra at various temperatures and observe the temperature-dependent spectral changes.

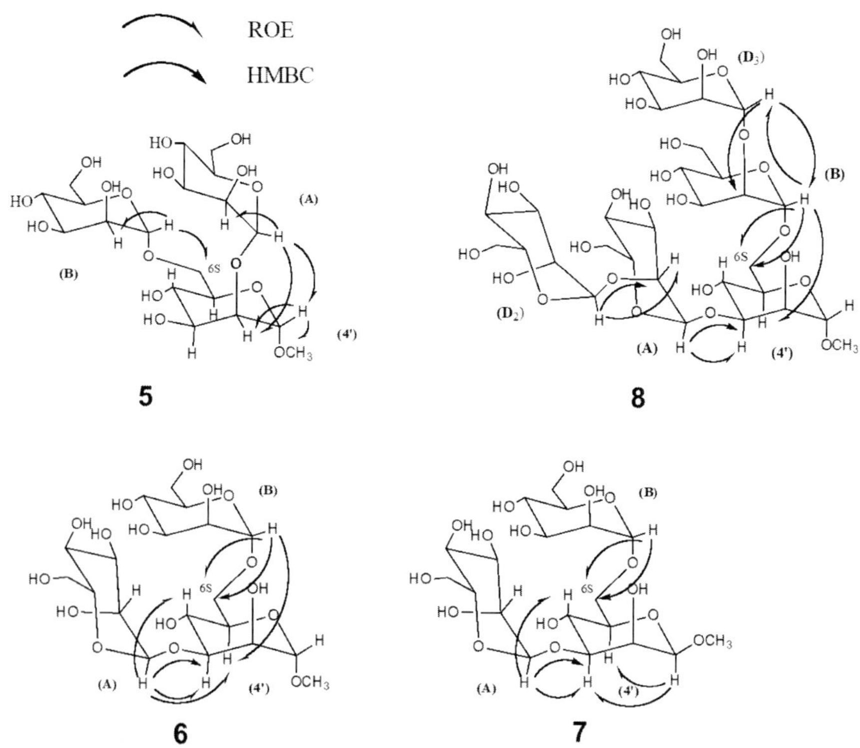

Figure 3. Interresidue ROE and HMBC connectivities identified in **5**-**8**.

Table 1. Summary of ^{1}H and ^{13}C chemical shifts of each mannose residue in 1-8. ^{1}H and ^{13}C chemical shifts are reported relative to DSS (4,4-dimethyl-4-silapentane-1-sulfonic acid) at 25°C

		1		2		3		4		5		6		7		8	
		^{13}C	^{1}H	^{13}C	^{1}H	^{13}C	^{1}H	^{13}C	^{1}H	^{13}C	^{1}H	^{13}C	^{1}H	^{13}C	^{1}H	^{13}C	^{1}H
4'	1	96.3	5.17	96.7	4.89	103.5	4.76	103.7	4.73	102.1	4.97	103.7	4.73	104.0	4.59	103.7	4.72
	2	73.4	3.92	73.9	3.93	72.6	3.93	72.6	3.92	81.4	3.95	72.2	4.09	72.6	4.15	72.2	4.07
	3	72.9	3.83	75.7	3.64	73.2	3.76	73.3	3.72	73.0	3.86	81.3	3.86	83.4	3.74	81.4	3.83
	4	69.5	3.66	69.3	3.57	69.5	3.63	69.2	3.73	69.2	3.80	68.3	3.88	68.5	3.81	68.4	3.86
	5	75.1	3.81	78.8	3.37	75.2	3.61	73.4	3.72	73.7	3.72	73.5	3.80	76.8	3.56	73.6	3.77
	6	63.7	3.76, 3.87	63.7	3.72, 3.90	63.6	3.76, 3.90	68.3	3.75, 3.95	67.9	3.76, 3.99	67.9	3.74, 4.01	68.1	3.79, 3.98	68.1	3.71, 3.98
A	1									105.0	5.02	105.1	5.10	105.1	5.11	103.5	5.32
	2									72.7	4.07	72.7	4.07	72.8	4.07	81.1	4.09
	3									73.1	3.80	73.1	3.88	73.1	3.90	72.8	3.98
	4									69.4	3.65	69.4	3.67	69.4	3.66	69.6	3.69
	5									76.0	3.73	76.0	3.74	76.0	3.80	76.0	3.74
	6									63.6	3.77, 3.89	63.6	3.77, 3.89	63.6	3.76, 3.90	63.8	3.72, 3.89
B	1							102.1	4.89	102.2	4.92	102.1	4.91	102.1	4.92	100.6	5.13
	2							72.7	3.96	72.7	3.98	72.6	4.00	72.6	4.00	81.3	4.01
	3							73.3	3.82	73.3	3.82	73.3	3.88	73.3	3.84	73.0	3.94
	4							69.4	3.65	69.3	3.68	39.4	3.67	69.5	3.68	69.5	3.71
	5							75.4	3.69	75.5	3.68	75.4	3.68	75.4	3.69	75.4	3.68
	6							63.6	3.75, 3.87	63.7	3.72, 3.89	63.6	3.77, 3.89	63.6	3.76, 3.90	63.6	3.76, 3.87
D2	1															104.9	5.04
	2															72.7	4.06
	3															73.0	3.84
	4															69.4	3.64
	5															75.9	3.73
	6															63.6	3.73, 3.84
D3	1															105.0	5.04
	2															72.7	4.06
	3															73.0	3.83
	4															69.4	3.64
	5															75.9	3.73
	6															63.6	3.73, 3.84
OCH_3						57.4	3.41	57.4	3.38	57.6	3.39	57.5	3.41	59.5	3.53	57.5	3.39

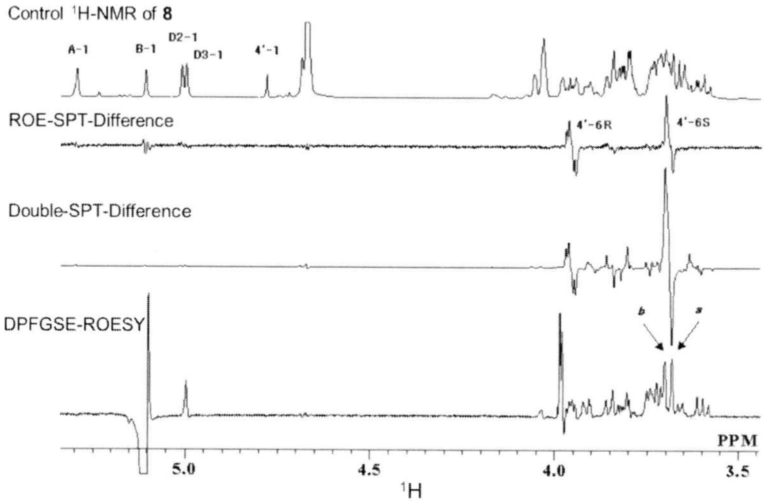

Figure 4. DPFGSE-ROESY, Double-SPT-Difference and ROE-SPT-Difference of **8**. In the ROE-SPT-Difference spectrum, Man-4' H6 protons are selectively observed. Control ^{1}H-NMR spectrum of **8** is shown at the top with assignment of anomeric signals.

It is best to selectively observe a limited set of signals. Here we follow the H6 protons in **8** (Figure 4). By irradiating Man-B H1 in the DPFGSE-ROESY sequence, we see the peak (arrow a and arrow b) originating from Man-4' H6proS (Kubota et al. 2012). Successive use of this peak for SPT (ROE-SPT difference), allowed signals from Man-4' H6proR and H6proS to be clearly differenciated from other signals. It should be noted that this NMR spectrum was collected over a very short time, and comprised only 128 scans. Man-4' H6proR and H6proS signals were then used as a probe for monitoring temperature dependency by obtaining DPFGSE-ROE-SPT difference spectra at 35, 45 and 55°C (Figure 5). Interestingly chemical shift changes were observed for both H6 protons, suggesting that this Man-B α1-6 Man-4' glycosidic linkage is in an equilibrium of several conformations. Furthermore, the coupling constant between H6proR and H5, ranged from 4.1 to 4.7 Hz. This suggests that the ω angle, i.e., the O1-C6-C5-O5 torsion angle, is dependent on temperature and the conformations in dynamic equilibrium.

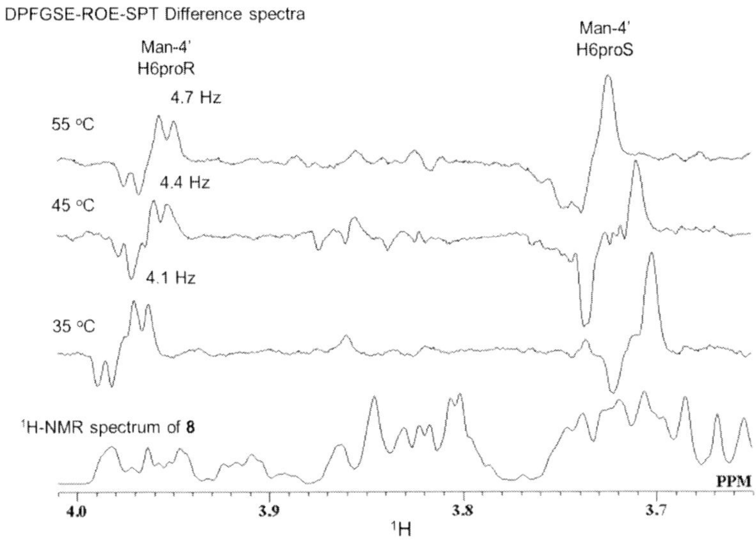

Figure 5. Temperature-dependent NMR spectral change of **8**. ROE-SPT-Difference was measured for the selective detection of Man-4' H6 protons as shown in Figure 4. Control ^{1}H-NMR spectrum of **8** is shown below.

NMR relaxation parameters ^{13}C T_1, T_2 and ^{1}H-^{13}C heteronuclear NOE can be used to probe oligomer dynamics. Relaxation of a nucleus is dependent on local magnetic fields generated by neighboring nuclei. The local magnetic field fluctuates because of the motion of the neighboring nuclei. Therefore, relaxation parameters such as T_1 and T_2 directly indicate the mobility of the nuclei. Such parameters are generally handled without reference to a specific model, but rather qualitatively discussed. Several papers have reported using ^{13}C relaxation parameters for analyzing carbohydrate dynamics with or without ^{13}C enrichment (Kjellberg et al. 1998, Yamaguchi et al. 1998, Zerbetto et al. 2018).

5. Oligomannose-Lectin Interaction

How then is the oligomannose structure recognized by a specific lectin? OS-9, one of the intracellular lectins, recognizes misfolded glycoproteins via a high mannose glycan (Hosokawa, Kamiya and Kato 2010). The trimannose unit Manα1-6Manα1-6Man of the high-mannose glycan is the recognition determinant. If OS-9 recognises this trimannose unit on a glycoprotein, it means that it is misfolded and it is then subjected to proteasomal degradation.

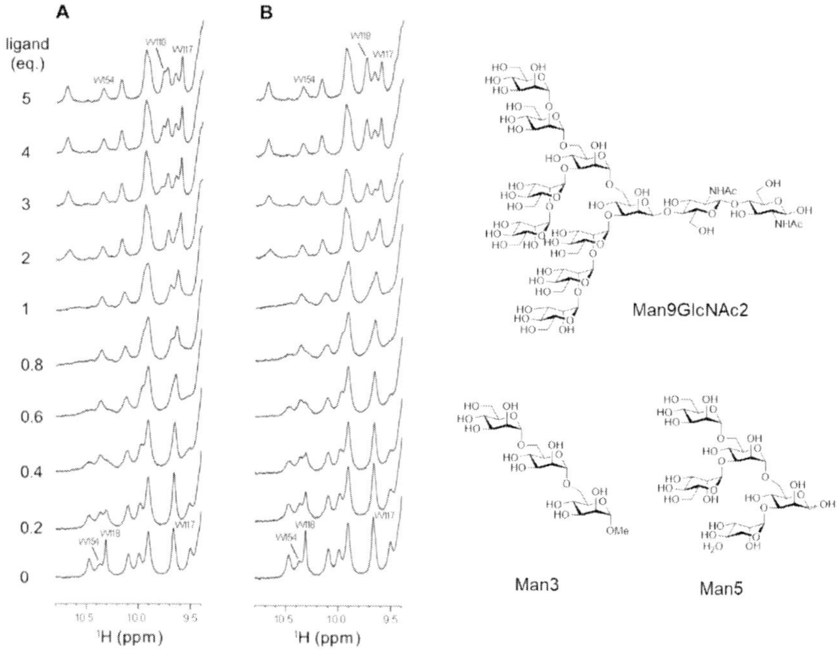

Figure 6. ¹H-NMR spectral changes of OS-9 upon addition of oligosaccharide ligands, Man3 (A) and Man5 (B). Indole NH-signals belonging to W117, W118, and W154 are indicated. OS-9 lectin domain was prepared in 0.44 mM solution, and oligosaccharide ligands were successively added from 0 to 5 equivalents. All NMR spectra were collected at 293 K. Chemical structures of Man9GlcNAc2, Man3 and Man5 are shown in the right.

The lectin domain of OS-9 was recombinantly expressed in *E. coli* and the protein in inclusion bodies isolated and refolded to a functional form. The crystal structure of the OS-9 lectin domain in complex with oligomannose (Man5) was determined at 2.1 Å resolution (Satoh et al. 2010).

In the ligand-binding site of OS-9, two mannose residues are clearly visible in the electron density map, which is assigned to the Man(B)-Man(4') unit. Interestingly, the indole ring of Trp118 is inserted into the α1,6 linkage region connecting Man(B) and Man(4'), which is stabilized through hydrophilic and hydrophobic interactions.

The electron density from Man(3) was not interpreted due to crystal packing problems but we thought it likely that Trp117 interacts with Man(3).

To demonstrate the putative interaction between Man(3) and Trp117, we performed a solution NMR analysis focusing on the Trp residues. First we conducted titration experiments in one-dimensional ^{1}H-NMR monitoring Man3 (Manα1-6Manα1-6Manα-OMe) and Man5 (Manα1-6[Manα1-3]Manα1-6[Manα1-3]Man) (Figure 6). The binding processes are evidently in fast exchange according to the chemical shifts. Dissociation constants were estimated to be 160 μM and 200 μM for Man3 and Man5, respectively. Their closeness is consistent with the fact that Man3 is the main recognition determinant of the OS-9 lectin.

NMR experiments on OS-9 reveal a more detailed picture of the interaction. Trp117 and Trp118 signals were successfully assigned using a selectively (^{13}C and ^{15}N) Trp-labeled OS-9 sample (Nishida et al. 2006).

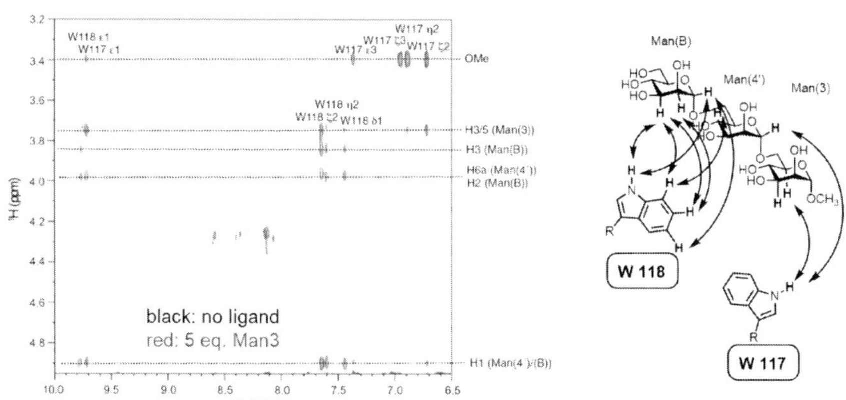

Figure 7. Overlay of the 2D ^{1}H-^{1}H NOESY spectra of OS-9 (black) with Man3 (red), focusing on the amide/aromatic region (A). The sets of strong transferred NOE signals indicate significant glycan-protein interaction under these conditions. Furthermore, intermolecular NOE signals are observed between Man3 and Trp117/Trp118. The intermolecular NOE correlations are summarized (B). Double-headed arrows indicate NOE correlations between the protons of the Man3 and the Trp residues of OS-9.

In the ^{1}H-^{1}H NOESY spectrum of OS-9 in the presence of Man5, intermolecular NOEs between Man5 and OS-9 were observed (Figure 7), which support our hypothesis that Man(3) stacks onto the indole side chain of Trp117. The Man(3) residue has a hydrophic patch formed by H1, H3 and H5 and these protons may be involved in a CH-π interaction with the Trp117 ring. CH-π interactions of this kind is often observed in carbohydrate-protein interactions (Weis and Drickamer 1996, Nishio 2011, Hudson et al. 2015, Spiwok 2017). In summary, we now know the mechanism whereby OS-9 specifically recognizes the Manα1-6Manα1-6Man unit in high-mannose type glycans through the double tryptophan motif. X-ray crystallographic analysis and solution NMR complement each other and a multifaceted approach is often essential for un-biased and comprehensive understanding of such biological events.

REFERENCES

Hori, H. 1987. "*Thesis, University of Tohoku*".

Hori, H., Y. Nishida, H. Ohrui, H. Meguro and J. Uzawa. 1988. "Conformational-analyses of α(1-6) and β(1-6) mannodisaccharides by deuterium substitution effect on relaxation rate and NOE". *Tetrahedr. Lett.*, 29 (35):4457 - 4460. doi: Doi 10.1016/S0040-4039(00)80521-3.

Hosokawa, N., Y. Kamiya and K. Kato. 2010. "The role of MRH domain-containing lectins in ERAD". *Glycobiol.*, 20 (6):651 - 60. doi: 10.1093/glycob/cwq013.

Hudson, K. L., G. J. Bartlett, R. C. Diehl, J. Agirre, T. Gallagher, L. L. Kiessling and D. N. Woolfson. 2015. "Carbohydrate-aromatic interactions in proteins". *J. Am. Chem. Soc.*, 137 (48):15152 - 60. doi: 10.1021/jacs.5b08424.

Kjellberg, A., T. Rundlof, J. Kowalewski and G. Widmalm. 1998. "Motional properties of two vicinally disubstituted trisaccharides as studied by multiple-field carbon-13 NMR relaxation". *J. Phys. Chem. B,* 102 (6):1013 - 1020. doi: DOI 10.1021/ jp971542e.

Kubota, Y., H. Hori, R. Sawa, H. Seki and J. Uzawa. 2012. "Structural analyses of mannose pentasaccharide of high mannose type oligosaccharides by 1D and 2D NMR spectroscopy". *Magn. Reson. Chem.,* 50 (10):659 - 64. doi: 10.1002/mrc.3859.

Nagae, M. and Y. Yamaguchi. 2015. "Sugar recognition and protein-protein interaction of mammalian lectins conferring diverse functions". *Curr. Opin. Struct. Biol.,* 34:108 - 15. doi: 10.1016/j.sbi.2015.08.005.

Nishida, N., F. Motojima, M. Idota, H. Fujikawa, M. Yoshida, I. Shimada and K. Kato. 2006. "Probing dynamics and conformational change of the GroEL-GroES complex by ^{13}C NMR spectroscopy". *J. Biochem.,* 140 (4):591 - 8. doi: 10.1093/jb/mvj188.

Nishida, Y., H. Hori, H. Ohrui, H. Meguro, J. Uzawa, D. Reimer, V. Sinnwell and H. Paulsen. 1988. "Conformational-Analyses of *O*-4, *O*-6-branching tri-D-glucopyranosides - Influence of *O*-4 linked residues on solution conformations about C5-C6 bonds at (1-6)-linkages". *Tetrahedr. Lett.,* 29 (35):4461 - 4464. doi: Doi 10.1016/S0040-4039(00)80522-5.

Nishio, M. 2011. "The CH/π hydrogen bond in chemistry. Conformation, supramolecules, optical resolution and interactions involving carbohydrates". *Phys. Chem. Chem. Phys.,* 13 (31):13873 - 900. doi: 10.1039/c1cp20404a.

Ogawa, T., K. Katano and M. Matsui. 1978. "Regio-controlled and stereo-controlled synthesis of core oligosaccharides of glycopeptides". *Carbohydr. Res.,* 64 (Jul):C3-C9. doi: Doi 10.1016/S0008-6215(00) 83724-6.

Ogawa, T., K. Katano, K. Sasajima and M. Matsui. 1981. "Synthetic studies on cell-surface glycans .3. branching pentasaccharides of glycoprotein". *Tetrahedr.,* 37 (16):2779 - 2786. doi: Doi 10.1016/ S0040-4020(01) 92345-2.

Ohrui, H., Y. Nishida, H. Itoh and H. Meguro. 1991. "Preferred conformation about the C5-C6 bond of *N*-acetylneuraminyl(2-6)-D-galactopyranosides and *N*-acetylneuraminyl(2-6)-D-glucopyranosides in solution". *J. Org. Chem.,* 56 (5):1726 - 1731. doi: DOI 10.1021/ jo00005a015.

Ohrui, H., Y. Nishida, M. Watanabe, H. Hori and H. Meguro. 1985. "^{1}H-NMR studies on (6R)-deuterated and (6S)-deuterated (1-6)-linked disaccharides - Assignment of the preferred rotamers about C5-C6 bond of (1-6)-disaccharides in solution". *Tetrahedr. Lett.,* 26 (27):3251 - 3254. doi: Doi 10.1016/S0040-4039(00)98164-4.

Rademacher, T. W., R. B. Parekh and R. A. Dwek. 1988. "Glycobiology". *Annu. Rev. Biochem.,* 57:785 - 838. doi: 10.1146/ annurev.bi.57.070188.004033.

Rudd, P. M. and R. A. Dwek. 1997. "Glycosylation: heterogeneity and the 3D structure of proteins". *Crit. Rev. Biochem. Mol. Biol.,* 32 (1):1 - 100. doi: 10.3109/10409239709085144.

Satoh, T., Y. Chen, D. Hu, S. Hanashima, K. Yamamoto and Y. Yamaguchi. 2010. "Structural basis for oligosaccharide recognition of misfolded glycoproteins by OS-9 in ER-associated degradation". *Mol. Cell.,* 40 (6):905 - 16. doi: 10.1016/j.molcel.2010.11.017.

Spiwok, V. 2017. "CH/ π Interactions in carbohydrate recognition". *Molecules,* 22 (7). doi: 10.3390/molecules22071038.

Suga, A., M. Nagae and Y. Yamaguchi. 2018. "Analysis of protein landscapes around *N*-glycosylation sites from the PDB repository for understanding the structural basis of *N*-glycoprotein processing and maturation". *Glycobiol.,* 28 (10):774 - 785. doi: 10.1093/glycob/cwy059.

Uzawa, J., Y. Fujimoto and S. Yoshida. 2006. "Incorporation of selective population transfer and homo-spin decoupling into selective one-dimensional experiments". *Magn. Reson. Chem.,* 44 (1):45 - 53. doi: 10.1002/mrc.1722.

Uzawa, J. and S. Yoshida. 2004. "A new selective population transfer experiment using a double pulsed field gradient spin-echo". *Magn. Reson. Chem.,* 42 (12):1046 - 8. doi: 10.1002/mrc.1495.

Varki, A. 1993. "Biological roles of oligosaccharides: all of the theories are correct". *Glycobiol.,* 3 (2):97 - 130. doi: 10.1093/glycob/3.2.97.

Varki, A. 2017. "Biological roles of glycans". *Glycobiol.,* 27 (1):3 - 49. doi: 10.1093/glycob/cww086.

Weis, W. I. and K. Drickamer. 1996. "Structural basis of lectin-carbohydrate recognition". *Annu. Rev. Biochem.,* 65:441 - 73. doi: 10.1146/ annurev. bi.65.070196.002301.

Winnik, F. M., J. R. Brisson, J. P. Carver and J. J. Krepinsky. 1982. "Syntheses of model oligosaccharides of biological significance - Synthesis of methyl 3,6-di-O-(α -D-mannopyranosyl)- α-D-manno pyranoside and the corresponding mannobiosides". *Carbohydr. Res.,* 103 (1):15 - 28. doi: Doi 10.1016/S0008-6215(82)80004-9.

Yamaguchi, Y., K. Kato, M. Shindo, S. Aoki, K. Furusho, K. Koga, N. Takahashi, Y. Arata and I. Shimada. 1998. "Dynamics of the carbohydrate chains attached to the Fc portion of immunoglobulin G as studied by NMR spectroscopy assisted by selective ^{13}C labeling of the glycans". *J. Biomol. NMR,* 12 (3):385 - 94. doi: 10.1023/A:100839 2229694

Zerbetto, M., T. Angles d'Ortoli, A. Polimeno and G. Widmalm. 2018. "Differential dynamics at glycosidic linkages of an oligosaccharide as revealed by ^{13}C NMR spin relaxation and stochastic modeling". *J. Phys. Chem. B,* 122 (8):2287 - 2294. doi: 10.1021/ acs.jpcb.7b12478.

In: Advances in Chemistry Research
Editor: James C. Taylor
ISBN: 978-1-53615-828-1

Chapter 2

PEGYLATION OF α-INTERFERON: BIOCHEMICAL, CLINICAL, ANALYTICAL AND STABILITY ASPECTS

Adeel Arsalan[1,*], Mirza Tasawer Baig[1], Kiran Qadeer[2], Muhammad Ibrahim Sial[1], Iqbal Ahmad[2] and Syed Abid Ali[3]

[1]Faculty of Pharmacy, Ziauddin University, 4/B, Shahrah-e-Ghalib, Clifton, Karachi, Pakistan

[2]Baqai Institute of Pharmaceutical Sciences, Baqai Medical University, Super Highway, Karachi, Pakistan

[3]H.E.J. Research Institute of Chemistry, ICCBS, University of Karachi, Karachi, Pakistan

ABSTRACT

Polyethylene glycol (PEG) is a versatile polymer not only used as an excipient in different dosage forms but it also plays an important role in the *in-vitro* and *in-vivo* stability of protein and polypeptide drugs. PEGylation helps to produce non-immunogenic, longer absorption and

* Corresponding Author's E-mail: adeelarsalan@hotmail.com.

cost effective therapeutic proteins mainly indicated in cancer, hepatitis C, Crohn's disease, immunodeficiency disease, leukemia, and rheumatoid arthritis. Interferon (IFN) is a natural anti-viral agent. It stands second position after somatropin in worldwide prescription drug sales and investment and developmental strategies in the global biopharmaceutical market. In this chapter the history, clinical use, and pharmacokinetics profile of PEG-IFN has been presented. Important features regarding its clinical efficacy in combination with new antiviral agents, like ribavirin, telaprevir, entecavir, tenofovir, adefovir, sofosbuvir and ledipasvir especially in hepatitis is also discussed.

The known and possible sites of PEGylation of IFN are pointed out. The use of chromatographic techniques like, ion-exchange chromatography (IEC), hydrophobic interaction chromatography (HIC), size exclusion chromatography (SEC), and/or high performance liquid chromatography (HPLC) are mainly used for separation and purification of PEG-IFNs. Moreover, the novel site specific PEGylation with novel PEG derivatives are also described. The analytical techniques like reversed phase (RP) HPLC, gel electrophoresis, mass spectrometry are mainly used for quantitative and qualitative estimation of IFN and PEG-IFN. The stability profile of IFN and PEG-IFN is also discussed. Finally the chapter also through light on the future prospects on other PEGylated therapeutics proteins.

1. Introduction

The formulation scientists in pharmaceutical industries are facing problems in the delivery of drugs. A number of excipients are used in the formulations to develop stable and effective medicines. In this effort low water solubility, lack of targeted action and enzymatic and pH degradation of drugs are mainly encountered. Major aim of the formulation development is to maintain the safety, efficacy and stability profile of drugs. The use of polymers to increase the solubility along with safety is a challenge for effective drug delivery system (DDSs). The conjugation of polymers with drugs plays an important role in DDSs. The polymers covalently bond with drugs and form a polymer-drug complex. This forms a shield and produces an umbrella-like effect to maintain the integrity of drugs (Kamaly et al., 2016).

The idea of bioconjugation of polymers with proteins leads to a new era of therapeutic protein drug delivery systems, mainly with the use of biocompatible and biodegradable polymers like poly(D,L-lactic-co-glycolic acid) (PLGA), poly(ethylene glycol) (PEG), polystyrene, poly (oligoethylene glycol methyl ether methacrylate) (POEGMA), poly(N,N'-diethyl acrylamide) (PDEAM), poly(N-(2-hydroxypropyl) methacrylamide) (PHPMA), poly(glutamic acid) (PGA) and poly(N-isopropyl acrylamide) (PNIPAM) (Damodaran and Fee, 2010).

Polyethylene glycol (PEG) is a versatile polymer also known as polyethylene oxide (PEO) or polyoxyethylene (POE) (Jang et al., 2015). The molecular size of PEG ranges from 300 to 100 kg/mol. The PEG with a molecular weight below 1000 g/mol usually exists in a liquid state. The increase in size of polymer changes into a waxy (more than 1k to less than 3k) semisolid to waxy solid (3 to 40k). It is neutral in pH and has been used as viscosity enhancer and suspending agent in pharmaceuticals. It is coupled with hydrophobic molecules to produce non-ionic surfactants. It is a biocompatible, non-toxic, inert, and hydrophilic moiety, which is easily eliminated from the body with any structural changes depending upon its molecular weight (D'souza and Shegokar, 2016).

PEG has been commonly used in emulsions, creams, ointments, gels, lotions and cosmeceuticals, like toothpaste, lipstick, deodorant, shaving foam and foundation. It has also been used as laxative and PEG suppositories are commonly used in constipation. It is also used as a lubricant in eye drops. PEG is also used as a co-solvent in oral liquid preparations. It is also added as a tablet binder, lubricant and for elasticity in soft gelatin capsules and in the coating of tablets (Jang et al., 2015).

PEG is highly soluble in water and soluble in organic solvents like alcohols, esters, glycols, ketones, aromatic hydrocarbons, and nitroalkanes, as well as soluble in benzene and dichloromethane. Its solubility depends upon molecular weight and temperature (Winger et al., 2009). Due to its versatility and solubility in water, it is mainly used as a protectant of pro-drugs (including both proteins and other biopharmaceutical drugs) (Banerjee et al., 2012).

1.1. Properties of PEG

PEG is a linear or branched polyether terminated with hydroxyl groups. It is synthesized by anionic polymerization of ethylene oxide initiated by nucleophilic attack of a hydroxide ion on the epoxide ring. It is mostly used for polypeptide modification by methoxy-PEG (mPEG) which is synthesized by anionic ring opening polymerization initiated by methoxide ions. The success of conjugation of PEG with biological products depends upon the chemical structure, molecular weight, steric hindrance, and the reactivity of the biological products. The bioconjugation is initiated by reactive or functional group such as –COOH, –OH, –SH, or $-NH_2$. Therefore, the synthetic methodology to form a conjugate involves either protection or deprotection of these groups (Banerjee et al., 2012).

1.2. PEGylation

Veronese and Mero (2008) defined PEGylation is the modification of biological molecules by covalent conjugation with PEG molecule. PEGylation was first reported by Davies and Abuchowskyy in 1970s for albumin and catalase surface modification. The first protein based therapy was recombinant insulin introduced in 1982 (Baeshen et al., 2014). The proteins based therapy is now a significant tool in the treatment and prevention of disease and illness and around more than 130 therapeutic proteins and peptides are approved by Food and Drug Administration (FDA) for clinical use. Recently, the use of proteins, peptides, hormones, enzymes, and even DNA and RNA therapy as biopharmaceutical drugs is increased (Farhadi and Hashemian, 2018). The solubility, thermal stability, and susceptibility to proteolysis of proteins depend on the structure, size, and ionic charge of proteins. Factors like enzymatic degradation (proteases),temperature, ionic strength and pH may results in the loss of intermolecular bonding, change in surface chemistry of proteins, unfolding and degradation which leads to the loss of *in vivo* biological activity (Meyer et al., 2002; Krishnamurthy and Manning, 2002). Similarly, external factors, like oxidation, photooxidation, acidic buffer hydrolysis leads to inactivation of proteins (Manneberg et al., 1995; Dalsgaard and Larsen, 2009) and *in-vivo*, the clearance of therapeutic proteins is increased due to shorter half-life, enzymatic degradation and reactions with immune systems (Goldberg and Gomez-Orellana, 2003). PEGgylation surpasses these hurdles and allows for a low volume of dose with prolonged circulation times by fusion and crosslinking of polymers with proteins thus changing the oligomerization state, glycosylation and mutations of amino acid residues (Burnham, 1994; Marshall et al., 2003)

Polymer therapeutics includes polymer drugs, polymer conjugates, and polymeric micelles (Burnham, 1994). PEG is slight immunogenic at high molecular weight and toxicities are observed with low molecular weight PEGs. PEG facilitates the moiety to cross by endocytosis to cell membrane at a particular target site. Furthermore, it is also considered as one of safest polymer intended for internal use. The main advantages of PEGylation are the modification of a protein and peptide drug to produce non-antigenic, high flexibility and longer absorption and cost effectiveness. It aids in the protection of protein drugs by better masking, modification of electro-osmotic flow, low volume of distribution, sustained adsorption from the injection site, and by creating a shield-like effect protecting against proteolysis and improved formulation properties of the protein (Chavez et al., 2018). The PEGylation of protein drugs like enzymes, cytokines, antibodies and protein scaffolds produces useful therapeutic applications in cancer, Crohn's disease, hepatitis C, immunodeficiency disease, leukemia, and rheumatoid arthritis (Awwad et al., 2018).

It depends upon the protein and PEG molar ratio, time of reaction, pH and temperature. The process is mainly classified as solution phase batch process and on-column fed-batch process (Fee and Van Alstine, 2006). In solution phase, N-hydroxyl succinimide ester, derivative of PEG is used as a reagent. The PEGylation reaction is performed at a ratio of 1:3-5 for 8-16 h at 4-8°C in a pH range of 7-8. The drug is acidified, diluted, and then filtered to obtain the purified PEGylated drug by cation exchange column (Akbarzadehlaleh et al., 2016). The separation of PEGylated and non-PEGylated proteins is carried out by using chromatography, like size exclusion chromatography (SEC), ion-exchange chromatography (IEC), hydrophobic interaction chromatography (HIC) and membranes or aqueous two-phase systems. The conjugation of PEG and protein is affected by choice of functional groups (Fee and Van Alstine, 2011).

The main problem with polypeptide therapeutics are shorter half-life, increased adverse immune response, low solubility, less thermal stability, precipitation, proteolytic degradation, immunogenicity and antigenicity (Damodaran and Fee, 2010). PEGylation of the polypeptide drug improves its pharmacokinetic and pharmacodynamic profiles. It improves greater compliance; better patient care and improved quality of life by once weekly regimen and helps to overcome the side effects.

1.3. PEGylation Chemistry

PEG is a synthetic polymer of monomeric ethyleneoxide. The initiation of polymerization is achieved by methanol for monofunctional PEG or by water leading to bifunctional polymer chains. The conjugation of protein is preferred by monofunctional PEG. It aids in the linking of two proteins with separate polymer chain. The mPEG is usually used as initiator and functionalized for protein PEGylation. The unnecessary cross linkage is prevented by dihydroxyl PEG (bifunctional polymer). Thus, the separation of diols is attained by conversion of hydroxyl group to carboxylic acid by IEC (Roberts et al., 2002).

The conjugation of proteins is done by mPEG by its free hydroxyl end group (Banerjee et al., 2012). The *C-terminal* carboxylic acid and *N-terminal* amino acid are used as site specific for binding by aldehyde functional polymers (van Vught et al., 2014). Moreover, lysine (LYS), arginine (ARG), histidine (HIS), glutamic acid (GLU), tyrosine (TYR), aspartic acid (ASP) and serine (SER) have shown reactivity with PEG. Lysine exists as the most abundant amino acid in protein sequences (Cho et al., 2007). The main obstacle in PEGylation is the amplification of conjugation. The amino group of proteins likes LYS conjugate by alkylation or acylation. The alkylation yields positively charged amines, whereas, acylation formed neutral amides (Damodaran and Fee, (2010).

1.4. Factors Affecting on PEGylation

It is a major concerned for formulation to produce stable and effective PEGylated proteins with maximum yield. A number of factors are involved in the PEGylation of proteins. These factors mainly include the viscosity of solution, role of excipients and protein concentration in solution, PEG and protein ratio, pH, temperature, incubation period, PEG molecular weight, role of protein hydrodynamics and site specific PEGylation of proteins (Akbarzadehlaleh et al., 2016).

1.4.1. Effect of pH

The pH of the solution plays an important role in the PEGylation of proteins. The pH range from 6 to 8 is preferred to enhance the conjugation of PEG-protein. The conjugation reaction is also carried out at pH 5.0 to increase conjugation and stability of proteins. The charge development in amino acids is mainly developed in the pH range 6-8. The conjugation of PEG with proteins mainly forms mono and di-conjugated PEG-proteins. In a study, Naik et al. (2017) establish the importance of pH to increase the PEGylation of insulin. It has been observed that mono conjugated precursor at 29 LYS of chain B yield is increased at pH 10.2. The main probability of conjugation is due to the attachment for the oligomer to B29, is the p*K*a of the LYS (Naik et al., 2017).

Proteins containing more than two HISs needed to modify the reaction to produce mono-PEGylated species by controlling the pH and concentration of reactants (Dozier and Distefano, 2015). PEGylation of human growth hormone is selective for *N-terminal* amine group of proteins. The *N-terminus* of many proteins are conjugated selectively by PEG aldehyde. It has also been found that at pH 5.0 the conjugation of *N-terminal* amine group occurs due to protonation of all amino groups on lysine (Da Silva Freitas et al., 2013).

Meng et al. (2008) used an average of 3.5 and 6.0 PEG chains per HSA molecule by conjugating 10-fold molar excess of PEG- Phenyl-Isothiocyanate, (3 and 5kDa, respectively), for 0.5 mM HSA at pH 6.5 and 9.2. There is no significant change in the secondary structure of HSA by PEGylation. However, the size, viscosity and osmotic pressure of 4% HSA increases the hexa PEGlation at pH 9.2 and results in PEG-HSA as a potential candidate of plasma volume expander (Akbarzadehlaleh et al., 2016).

The PEG conjugation with proteins is more selective and specific by induction of PEG aldehyde capable of forming stable secondary amine linkage with amino groups through reductive alkylation using sodium cyanoborohydride. It occurs due to the reactivity of aldehyde groups based on the nucleophilicity of amine groups. The conjugation occurred if pH of the solution is near or above the p*Ka* of specific amine terminal group. While, the heterogeneity of the PEG product is markedly decreased by the pH of reacting solution (Damodaran and Fee, 2010).

In a mild acidic pH medium, aldehyde is used to react with protein by site-specific modification at their *N-terminus*. The pK_as of *N-terminal* α-amino group and ε-amino group on lysine are ~ 7.8 and 10.1, respectively. Whereas, an acidic pH amine group of lysine is mainly protonated and unable to conjugate with the aldehyde group, while making the free amine on the *N-terminus* the only site for modification (Dozier and Distefano, 2015).

1.4.2. Effect of Temperature

Bonding between the PEG and proteins also depends upon temperature. The conjugation of PEG-proteins is increased at a particular temperature (Damodaran and Fee, 2010). The conjugation activity based on geometric and electronic capability and buffer system and temperature plays a major role in the preparation of PEGylated proteins. The separation and purification of PEGylated proteins is carried out by different chromatographic techniques preferably at low temperature (4-6°C) and at a particular pH. At low temperature, reaction increases the yield of mono-conjugated PEG-proteins. It has been observed that by pH adjustment mono-conjugated PEG-protein is produces more at 2-8°C than that at room temperature (Naik et al., 2017).

It has been observed that the peptide self-assembles in helices at an acidic pH, but is converted into β-sheets and slows down in conjugated PEG-proteins at higher temperature (Top et al., 2011). Similarly, coiled-coil peptide structure is retained by conjugation to PEG at higher temperature (Vandermeulen et al., 2003).Thus, it reduces thermal induced unfolding at higher temperature. PEGylation of coiled coils confers greater stability against changes in different conditions (e.g., pH, temperature and concentration) without perturbing the secondary structure (Hamley, 2014).

1.4.3. Effect of PEG Moiety

The nature and size of the PEG moiety on interaction causes variation in the pharmacokinetics, pharmacodynamics, and dosing regimens of therapeutic proteins (Bruno et al., 2012). The concentration and molecular weight of PEG significantly affect the conjugation of proteins. It has been observed that by a decrease in concentration of oligomers against precursor for conjugation, there is an increase in the formation of mono-conjugated PEG-proteins at 2-8°C with pH adjustment. It has also been found to give maximum formation (i.e., 71.4%) of mono-conjugated PEG-proteins (Naik et al., 2017).

The physical stability of PEG-lysozyme is influenced by the molecular weight of PEG. Lysozyme is conjugated with 2, 5 and 10 kDa mPEG-aldehyde. It has been observed that lower molecular weight PEGs have higher residual activity by producing change in enzymatic activity. Whereas, higher molecular weight PEG increases colloidal properties and conformational stability more than 16-fold (Morgenstern et al., 2017).

Yang et al. (2011) has used different size of ethylene polymers to form bonding with amino group of 29 LYS B chain. The conjugation of PEG did not change the secondary structure of insulin and reduces the solvent accessible surface area. PEG excludes water from the surface of insulin due to hydrogen bonds and hydrophobic interactions. These effects are increased by increasing the chain length of PEG up to 4kDa size.

The size of PEG-proteins is increased as compared to the original size by increasing the molecular weight of PEG. In an earlier study it has been observed that trypsin (a proteolic enzyme) is conjugated with different size of PEG polymer (i.e., 2, 5, 10 and 20 nm), and there is an increase in the size and molecular weight compared to original trypsin. The PEGylation of trypsin increases from 49 to 60°C (T_d) and thus enhance the enzymatic activity. Quantitatively, this PEG-trypsin conjugated enzymatic activity is increased up to 221-432% at 37°C (Chiu et al., 2010). Wu et al. (2014) found that long chain PEG caused stronger interactions and stable binding with proteins. The interaction between PEG and proteins is determined by fluorimetry, atomic force microscopy, and nuclear magnetic resonance. The intensity of PEG-protein interactions is due to change from hydrophilicity to amphiphilicity of molecular weight of PEG. The increase in PEG-protein interactions is mainly due to modification in PEG from hyrophilicity to amphiphilicity with increases in molecular weight of PEG.

Similarly, dextranases, an enzyme, increases in size by conjugation of mPEG (5kDa). It has been observed that there is an increase in the molecular weight of dextranases determined by Superose 12 (23 ml) SEC column and is directly proportional to concentration of mPEG. The residual activity of partially PEGylated dextranase (mPEG 5000 of 5.8 mg/ml) was 33.8% and for the completely PEGylated dextranase (mPEG 5000 of 29 mg/ml) it was 25.75%. The PEGylation of dextranases is decreased by increase in PEG polymer size, at an mPEG concentration of 5.8 mg/ml. At a concentration of 29 mg/ml several PEGylated peaks were produced with subtle difference in dextranase activity toward dextran T500. Retardation in the activity with increasing molecular weight clearly appeared with Sephadex G75 but for Sephadex G200 the retardation was a little slow (Abdel-Aziz and Janson, 2012).

The *p*-aminophenylanaline and lysine conjugation with different PEG units (4, 8, 12 kDa) into streptavidan was achieved using a frame shift suppression method based on a four condon anticondon pair (Abdel-Aziz and Janson, 2012). It creates a passage to larger PEGs (20-40 kDa) which are more applicable in therapeutic use.

PEGylation of human serum albumin (HSA) at the *N-terminal* end by using mPEG-propionaldehyde with molar ratio of (4:1) in sodium acetate buffer (100 mM, pH 5.0) for 20 h at room temperature has been carried out. The unconjugated and PEG-HAS conjugate was separated by hydrophilized polyvinylidene and hydrophobicity was induced by lyotropic salt that caused phase transition of PEG to the hydrophobic form. This resulted in the first elution of mono-PEGylated HAS at low concentration of salt (Akbarzadehlaleh et al., 2016).

1.4.4. Effect of Site Specific PEGylation

The conjugation reaction of PEG with proteins has limited chemo selectivity. Amino acids (LYSand CYS) on the surface of protein were non-specifically modified with various PEG electrophiles like succinimides, maleimides, chlorotriazines and aldehydes that differed in location as well as the number and occupancy of PEGylation sites. The development of new bioorthogonal molecules by unnatural amino acids bearing orthogonally reactive functional groups (i.e., azides, alkynes, ketones, and alkenes) allows good control of the site and degree of PEGylation. These unnatural side chains may selectively be modified with a suitable PEG reagent to form a PEGylated protein. In a study, alkyne or azide functionalized side-chain has been formed using an azide or alkyne PEG, respectively, and copper-catalyzed alkynezide/alkyne cycloaddition to give a triazole-linked PEG-protein conjugate. Similarly, a ketone functionalized side-chain has been attached to PEG alkoxyamine to form an oxime-linked PEG-protein conjugate (Lawrence and Price, 2016).

PEG aldehyde has shown selectivity to *N-terminal* α-amino group of a protein by reductive alkylation, due to the difference in p*K*a between the *N-terminal* 7agr;-ammonium and LYS ε-ammonium groups. CYS is one of the major amino acid residue of targeted selectivity mainly because of the high nucleophilicity of Cys thiol, for attachment to PEG vinyl sulfones and PEG maleimides; disulfide formation with PEG ortho-pyridyl and nucleophilic substitution with PEG iodoacetamides (Lawrence and Price, 2016).

It has also been reported that the size of PEG molecule and the site of PEGylation of IFN-α may lower the *in-vitro* activity Grace et al. (2005) have formulated monoPEGylated IFN-α-2b by using the linear PEG molecule with varying molecular weight of 5-30 kDa attached at 7 primary reactive sites like CYS^{1}, HIS^{34}, LYS^{31}, LYS^{83}, LYS^{121}, LYS^{131} and LYS^{134}. Moreover, the antiviral and antiproliferative activity has been evaluated by STAT translocation and HIS^{34} and CYS^{1} show the highest and the lowest residual activity, respectively. While, lysine residues possess intermediate activity in the order of LYS^{134}> LYS^{83}~ LYS^{131}~ LYS^{121}> LYS^{31}. It has also been observed that the size of PEG molecule may interfere with the interaction and binding of IFN-α to the IFNAR1-IFNAR2 heterodimeric receptor.

1.4.5. Effect of Incubation Period

The dextranase can be PEGylated on 10, 20, 30, 60 and 180 mins incubation period. It has been found by using sodium dodecyl sulfate-polyacrylamide gel electrophoresis (SDS-PAGE) that PEGylation is enhanced by increasing the concentration of mPEG and the incubation period. Suo et al. (2009) has PEGylated HSA by adsorbing it on an anion-exchange chromatographic media at pH 7.0 (10 mM PBS buffer) for 4 h at room temperature. A linear gradient elution was demonstrated over 60 min. Kouchakzadeh et al. (2010) incubated 5 mg BSA nanoparticles with 32.5 g/lit of PEG at 27°C for 10 min in phosphate buffer pH 7.0 to yield PEGylated nanoparticles with a mean diameter and zeta potential of of 217 nm and -14 mV, respectively.

The PEGylation of recombinant human granulocyte colony-stimulating factor (rh-GSF) has been carried out at room temperature (pH 5.0) by methoxy polyethylene glycol propionaldehydes with molecular weights of 10, 20, and 30 kDa at a ratio of 5:1. It has produced a ≥ 95% pure material with 86% mono PEGylated protein and showed 20.8-42.73% bioactivities (Behi et al., 2018).

2. INTERFERONS (IFNS)

Interferon (IFN) is the defensive response of the body after viral invasion. It was discovered by Isaacs and Lindemann (1957) as an antiviral factor produced by viral infected chick cells (Samuel, 2001). It is a member of cytokine family of proteins and rarely glycoproteins and secreted by somatic cells. It also aids in the activation of immune system like macrophages and B-cells. Its main functions are to kill infected cells by T-cells and promote apoptosis and antitumor activity. IFNs also increase the ability of resistance against new viral infection on the uninfected host. These possess antiviral activity against hepatitis B and C virus, herpes simplex virus, cytomegalovirus, vesicular stomatis virus, human immunodeficiency virus, polio virus, rhino virus and adeno virus. IFNs are involved in several immune reactions like inducers, regulators, and effectors during viral infection. It is soluble in water (Le Page et al., 2000; Samuel, 2001; Stetson and Medzhitov, 2006; Hou et al., 2009).

2.1. Types of IFNs

The family of IFNs is classified into four types. The IFNs present in human are IFN-α, IFN-β, IFN-γ. In human around 18 IFN-α non-allelic genes and a single IFN-β gene have been reported. Whereas, all human IFN-α elicited the similarly types of cellular responses, while the reported level of antiviral antiproliferative efficacies on various cells are quite different (Klaus et al., 1997). The molecular weight of IFNs ranges between 20 to 40 kDa (Metz, 1975; Joklik, 1977). The IFN-α (monomeric) type-1-IFN, has thirteen (13) subtypes mainly two IFN-α-2a and IFN-α-2b with more than 150 amino acids and molecular weight around 19 to 22 kDa. Quite high level of primary structural identity (i.e., 80%) have been reported between the various IFN-α subtypes, and around 50% sequence homology observed between these species and the single IFN-β subtype (Fig. 1). The three (3D) dimensional structure of human IFN-α-2a, elucidated by high resolution nuclear magnetic resonance (NMR)spectroscopy in solution, comprises of typical all alpha and four up-down helical bundle structure (Fig. 2a). The 3D structure is very similar to that of IFN-α-2b and murine IFN-β (Klaus et al., 1997).IFN-α-2a and IFN-α-2b are semi-synthesized in PEGylated form.

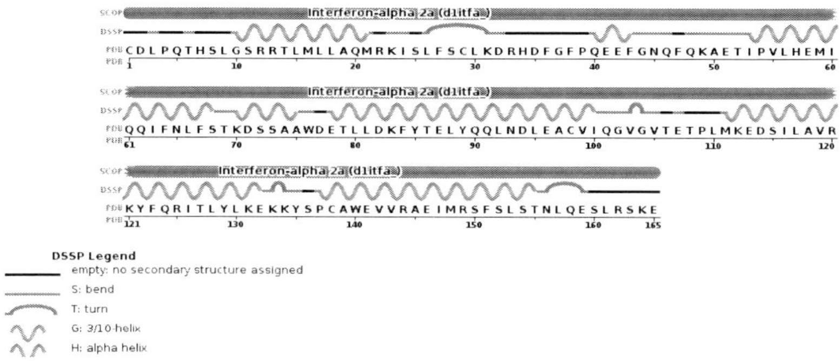

Figure. 1. Primary and secondary structure of human IFNα-2a illustrated in order to see protein sequence with reference to residual position for SS Bond (- - -), PEGylation (i.e., CYS, LYS, HIS, TYR etc.), receptor recognition (Δ) and secondary structural features (Figure is modified from Protein structure data bank, pdb i.d. code 1itfa).

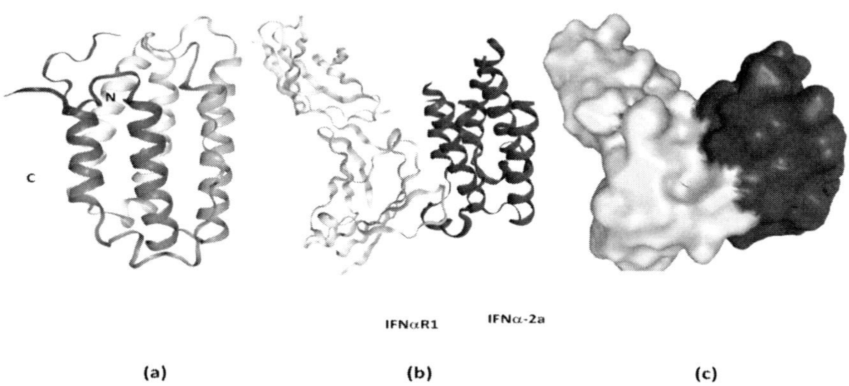

Figure. 2. Three (3D) dimensional structure of human IFNα-2a illustrating the typical all alpha and four up-down helical bundle structure (a), 3D IFNα-2a - receptor (IFNαR1) complex (b) and interaction surfaces (c). Figures are generated using the program (NGL Viewer). The 3D coordinates (pdb i.d. codes: 1itfa and 3s9d, a and b-c, respectively) were obtained from Protein structure data bank (www.rcsb.org).

On the other hand, monomeric IFN-β, has three (3) subtypes and belongs to type-1-IFN having 166 amino acids with a molecular weight of 18 to 25 kDa. In contrast, IFN-ω comprises 172 amino acids and approximate molecular weight of 20-25 kDa. All type-I-IFNs bind to the specific receptor complex known as IFN-α receptor i.e., IFNAR1 and IFNAR2 (Quadt-Akabayyov et al., 2006; Thomas et al., 2011; Krause et al., 2013) (Fig. 2b-c). While the IFN-γ belongs to type-II-IFN and binds to IFNGR receptor which is composed of 143 amino acids with molecular weight of 20-25 kDa. Lastly, IFN-λ 1, IFN-λ 2 and IFN-λ 3 belong to type III IFNs having 200 amino acids and bind with receptor CRF2-4and IFNLR1 also known as CRF2-12 (Kontsek, 1994; Gibbert et al., 2013; Paul et al., 2015; Tsuge et al., 2017).

2.2. Mode of Action

The IFNs exhibit activity by binding to specific receptors on cell membrane and initiate complex intracellular actions. They show significant cellular specificity for example, the IFN produced by mice cells is not active in human cells. Moreover, they are non-toxic to uninfected cells. There is induction of some enzymes and repression of cell proliferation and immuneregulatory responses like phagocytosis, macrophages, cytotoxicity of lymphocytes of target cells and inhibition of viral replication in infected cells. IFNs are synthesized and released by viral infection (PIL, 1997). They are not definite for a specific virus. IFN preparations also exhibit biological activity against tumor cell growth and immune response. It can be induced not only by virus but also by synthetic and natural polyribonucleotides. IFNs produce alterations in intracellular regulatory molecules to inhibit viral growth (Joklik, 1977). The invasion of virus on host cell results in infection by the release of nucleic acid which increases the production of new viruses. Meanwhile, IFN genes are turned on and form IFNs by mRNA. IFNs activate the production of antiviral proteins which produces its activity against viruses by inhibition of viral reproduction (Metz, 1975; Joklik, 1977). IFNs also induces the inhibition of viral proteins by protein kinase RNA-dependent, 2-5A synthetase (IFN inducer enzyme) that kills the virus by cleavage of its RNA and Mx proteins and block the activity of polymerase enzyme of the viruses.

2.3. Indications

The IFNs bind with different receptors and produce multiple intracellular regulatory molecules. The IFNs is indicated in different diseases like IFN-α in hairy cell leukemia, malignant melanoma, chronic myelogenous leukemia, AIDS related Kaposi's sarcoma, genital warts (condylomata acuminata) condylomata acuminate, follicular lymphoma, condylomata acuminata, and hepatitis B and C; IFN-β in multiple sclerosis and IFN-γ against chronic granulomatous disease (Stuart-Harris et al., 1992; Samuel, 2001).

2.4. Side Effects of IFNs

The major side effects observed by IFN are flu, fatigue, muscle and joint aches, loss of appetite, diarrhea and depression, whereas, metallic taste, dry skin, dry mouth, gastrointestinal upset, elevated liver function test, neurological complaints and cytopenia also appears. The prolonged use of IFN leads to thyroid and blood related problems, bleeding, hemolytic anemia, seizures, thrombocytopenia, and neutropenia. The level of bone marrow reverts to its normal state by stopping the treatment (Samuel, 2001; Ahad et al., 2004).

3. PEG-IFNs

Innovation in new drug delivery systems is the major effort for commercial and intellectual exploitation. Davis with his co-workers (1977) prepared first PEGylated bovine serum albumin and bovine liver catalase by covalently linked methoxyPEG (mPEG) (Mishra et al., 2017). In the last three decades, there has been a rapid growth in PEGylation of proteins, peptides, hormones and enzymes, liposomes and nanoparticles (Damodaran and Fee, 2010). Globally, two types of PEG-IFNs are commercially available for clinical use. These two types of α-IFNs are PEGylated-IFNs.

In the PEGylation of IFN with PEG, a branched 40 kDa and linear 12 kDa is conjugated by covalent bond with IFN-α-2a at LYS residues and IFN-α-2b via an unstable urethane bond that is hydrolyzed after injection, respectively (Bruno et al., 2012). There are significant difference in the chemistry, pharmacology, indications, and side effects of PEG-IFN-α-2a (Pegasys®) and PEG-IFN-α-2b (Pegintron®). These were launched in 2001 and 2002 by Schering Corp., Ireland and Hoffmann-La Roche, Inc., Switzerland, respectively (Foster, 2004; Zheng et al., 2015). The physical and chemical characteristics of PEG-IFN-α- 2a and 2b are presented in Table1.

3.1. Chemistry of PEGylation of IFNs- α

The PEGylation of proteins is performed in excess amount of viscous PEG solution with a high concentration of proteins (Jevsevar et al. 2010). The PEGylation of IFNs is an expensive procedure and to avoid the loss of drug it needs special techniques for the separation of PEG-IFN and non-PEG-IFN, mainly, the PEG conjugation with proteins with LYS and *N-terminal* amino group. PEG is hydrophilic inert biocompatible molecule is cross-linked with α-IFNs in different ways at different positions. The initial PEGylation is carried out through acylation by first generation activated mPEGs like, succinimidyl carbonate (SC-PEG) (Zalipsky et al. 1992; Miron and Wilchek, 1993) and benzotriazole carbonate (BTC-PEG) (Roberts et al., 2002). In SC and BTC, PEG mainly reacts with the LYS residues and form carbamate linkage and reacts with the HIS and TYR residues. It has been found that the hydrolytic degradation of BTC-PEG is slightly faster than the SC-PEG under same condition at pH 8.0 and 25°C (Roberts et al., 2002). SC and BTC form hydrolytic imidazolecarbamate linkage by coupling with HIS residues of α-IFN, suitable for controlled release preparations (Lee and McNemar, 1999). The molecular weight of unmodified α-IFN is 19 kDa, by the attachment of 40 kDa and 12 kDa PEG molecules, the cumulative size of α-IFN is increased 3 and 2-fold, correspondingly (Foster, 2004).

The conjugation of PEG with IFN-α-2a is different from that the PEG-IFN-α-2b. There is a covalent attachment of branched 40 kDa PEG with IFN-α-2a (Noureddin and Ghany, 2010). The process of PEGylation needs to design a method with an increase in PEG-biological product conjugation. Dicyclohexyl carbodiimide (DCC), 1-ethyl-3-(3-dimethylaminopropyl) carbodimide (EDC) or N-hydroxysuccinimide (NHS) esters are commonly used for conjugation (Banerjee et al., 2012).

The branched chain is attached with the side chain and *N-terminal* amine group of LYSs and then by activation of carboxylic acid a stable amide bond is formed. On the activation of carboxylic acid with NHS helps branched chain PEG to form stable amide bond. The monosubstituted conjugate of PEG-IFN-α-2a is synthesized by using a small amount of PEG as chain linker. Moreover, it shows stability against hydrolysis, and is dispensed in the solution form. This may bound to receptor with the intact PEG (Foster, 2004).

PEG-IFN-α 2a has superior activity as compared to PEG-IFN-α-2b and has a prolonged half-life. PEG-IFN-α-2a has lower rate of clearance from body than unmodified IFN-α-2a. The PEGylation of IFN-α-2a is achieved by LYS and HIS residues. The increase in size of conjugated PEG results in a substantial failure of therapeutic application by severe loss of bioactivity. The clinical use of PEG-IFN-α-2a produces a greater response rate as compared to the conventional IFN-α-2a, in liver cirrhosis. It exhibits sustained adsorption by reduced renal clearance with strong antiviral activity by once weekly dose (Noureddin and Ghany, 2010).

The conjugation of PEG with IFN is done by amide bond reaction between *N*-hydroxysuccinimide ester derivative of PEG molecule and the free amino group of IFN-α-2a. Several conjugation strategies are now available, such as alkylation, which maintains the positive charge of the starting amino group because a secondary amine is formed, or on acylation, accompanied by loss of charge. The site-specific amino modification aids in the purification of the PEGylated products (Banerjee et al., 2012).

LYS is the most prevalent amino acid in proteins. The modified multiple LYSs increase the chances of PEG for attached per protein molecule (PEGmers) but it can also be attached to different amine sites of protein. The main hurdle in PEGylation is non site-specificity and multiple attachments (interfere with the receptor binding sites). It can be overcome by site-specific PEGylation by reductive alkylation of the *N-terminus* amine, which is achieved by conjugating a methoxy PEG (mPEG) molecule under acidic condition. It occurs in the solution phase with three to four purification steps to remove undesired species, such as the reducing agent, unreacted PEG, and multi-PEGylates (Dozier and Distefano, 2015).

The monosubstituted conjugate of PEG with IFN-α-2b is also synthesized by the same process as that of IFN-α-2a (Noureddin and Ghany, 2010). The IFN-α-2b is cross-linked by urethane bond in which 12 kDa linear SC-PEG molecules are covalently conjugated primarily to HIS residue with the remaining PEGylation at LYS and the *N-terminal* CYS, TYR and SER residue (Foster, 2004; Grace et al., 2005). It is susceptible to hydrolytic degradation, and thus IFN-α-2b is separated from PEG and circulates as a prodrug after subcutaneous (s.c) administration (Wang et al., 2000; Grace et al., 2001). Due to separation from PEG, IFN-α-2b exerts its activity as a conventional IFN-α-2b with same pharmacokinetic profile (Foster, 2004).

3.2. Pharmacokinetics of PEG- α -IFNs

The PEGylation of α-IFNs structurally change to its original chemical entity. It acts to show an umbrella effect and produces shield like effect. The differences in basic chemistry between the conventional α-IFNs and PEG-IFNs produce variation in pharmacokinetics and pharmacodynamics profile and have significant change in dosing and patient management. The absorption and circulation of PEG-IFN-α-2b is rapid and undetectable within a few days, while, PEG-IFN-α-2a is slowly absorbed, well perfused in liver and detectable after a week. The elimination half-life of conventional IFN-α-2b in subcutaneously and intramuscularly administration is 2-3 h, but absorption half-life of PEG-IFN-α-2b is 4.6 h (Pedder, 2003).

Intramuscular (i.m) administration of PEG-IFN shows more than 80% bioavailability having a volume of distribution (V_d) from 12-40 L with 31L mean V_d or 0.4 L/kg in healthy volunteers. PEGylation of IFN decreases the clearance of IFN and increases the half-life. The total body clearance of IFN is nearly twice the glomerular filtration rate (Noureddin and Ghany, 2010; Zheng et al., 2015).

The intravenous (i.v) administration rapidly eliminates IFN from blood. Its V_d is ~ 20-60%.The clearance of IFNs family varies from 4.8 to 48 L/h, whereas, the terminal elimination half-lives of α-IFN and β-IFN are 4-6 and 1-2 hours, respectively, and 25-35 minutes for γ-IFN. There is a greater absorption of α-IFN and β-IFN (>80%) compared to γ-IFN (30-70%) after i.m and s.c administration (Wills, 1990). The clearance of PEG-IFN-α-2a is from the liver and kidney, whereas, ~ 30% of PEG-IFN-α-2b is eliminated by the kidney and the remaining is cleared by the liver or degraded after interacting with cellular IFN receptors (Zeuzem et al., 2003). PEG-IFN-α-2a has significantly reduced renal clearance compared to PEG-IFN-α-2b (Rajender et al., 2002). The pharmacokinetic profile of PEG-IFN-2a and 2b are presented in Table 1.

In a human study when IFN-α-2b was administered by i.v, i.m and s.c in 12 healthy volunteers. It was found that i.v infusion is rapidly eliminated from the serum with elimination and distribution phase half-lives of 1.7 and 0.1 h, respectively. However, s.c and i.m have shown similar absorption half-lives of 5.5 and 5.8, respectively. The C_{max} of s.c and i.m is 45.8 IU/mL at eight hours and 42.1 IU/mL at six hours, respectively. The elimination half-lives were 2.9 and 2.2 h for s.c and i.m, correspondingly. The area under serum concentration of i.v infusion is less than that of s.c and i.m (Radwanski et al., 1987). The V_d of PEG-IFN-α-2b is ~ 30% lower than that of the conventional IFN-α-2b (Zeuzem et al., 2003).

Rodríguez et al. (2000) conducted a randomized, double blind and two-compartment study of INTRON A and HEBERON ALFA R (recombinant IFN-α-2b) administered single 10 x 10^6 IU and i.m injection in 24 healthy volunteers. The authors observed that the peak serum concentration (C_{max}) were 15 and 18 IU/ml after 5 and 7 h time max plasma concentration (T_{max}) for INTRON A and HEBERON ALFA R, respectively. There is no significant variation in the clearance of these two recombinant IFNs, whereas, INTERON A showed inferiority against HEBERON ALFA R in

V_d, mean input time, mean residence time, half value duration and area under curve.

Table 1. Physical, chemical and pharmacokinetics characteristics of PEG-IFN-α-2a and 2b

Characteristics	**PEG-IFN-α-2a**	**PEG-IFN-α-2b**
PEG structure	Two 20kDa Peg chains, large, branched 40 kDa mPEG	Small, linear 12kDa monomethoxy (m) PEG
Molecular weight	60 kDa	31 kDa
Stability	Stored as a solution for up to 2 years	Hydrolytic degradation, Stored as a powder, reconstituted before injection
Positional isomer	6	13
PEG conjugation	mPEG activated by N-hydoxy-succinimide attached with LYS residues	mPEG activated by succinimidyl carbonate reacted with certain amino acids
PEG-protein Bond	Stable amide bond between mPEG and LYSs	Unstable urethane (carbonyl) bond
Absorption half-life	50 h	4.6 h
Peak serum concentration (C_{max})	72-96 h, sustained absorption	15-44 postdose and 48-72 sustained (10 fold greater)
Elimination half-life	40 h	22-60 (5 fold greater half-life)
Volume of distribution (V_d)	6-14 L/kg	0.99 L/kg
Clearance	100 ml/h	22 ml/h (seven fold lower)
Time max plasma concentration (T_{max})	80 h	20 h
Peak-to-through ratio	1.5-2.0	>10

(FDA, 2003; Zeuzem et al., 2003; Foster, 2004; Singal, 2010; BCCA, 2016)

The pharmacokinetic study of IFN-α-2b after intraperitoneal (i.p) 5, 10, and 15 million units (MU) administration in metastatic cancer showed that the bioavailability of i.p is 30-fold higher in peritoneal fluid than in the peripheral blood, thus reducing the incidence of side effects. The elimination half-life of i.p infused IFN-α-2b was 10-32 h. It has been observed that IP bolus distributed IFN from blood with half-life below 2 h and dose dependent C_{max} 47 IU for 3 MU and 145 IU for 5 MU. However, 24 h infusion leads to a steady state 4 to 6 h and maximum concentration of 8.5 or 12.5 IU/mL, respectively, with a half-life of 16 h (Schuller et al., 1992).

A single dose of PEG-IFN-α-2a reached C_{max} 14.2 μg/L in a mean interval of 78 h after administration (Noureddin and Ghany, 2010). The weekly multiple dose of PEG-IFN-α-2a achieved C_{max} 25.6 μg/L in the mean time of 45 h, but in the case of PEG-IFN-α-2b it reached C_{max} between 15-44 h after the administration. The smaller peak-to-trough ratio of PEG-IFN-α-2a exhibits less variation against PEG-IFN-α-2b for a week dosing (Zeuzem et al., 2003). It indicates that there is a limited need to adjust the dose of PEG-IFN-α-2a as compared to PEG-IFN-α 2b which must be adjusted according to the body weight or administered twice a week, otherwise it may leads to the rebound of virus (Foster, 2004; Di Bisceglie et al., 2007; Noureddin and Ghany, 2010).

4. Combination Therapy with PEG- α -IFNs

IFNs mainly used in hepatitis C, hepatocellular carcinoma (HCC), liver transplantation, hepatic failure, and hepatic related infections. It provides significant beneficial therapeutic effect in these infections. Mostly hepatic infected patients are unaware of their infection because of the slow appearance of symptoms (Lee et al., 2017). The drugs used in combination with IFNs are briefly summarize as follows.

4.1. Ribavirin (RBV)

Ribavirin (RBV) is an antiviral agent effective against both DNA and RNA virus. It was synthesized in 1970 and used to treat hepatitis C virus (HCV) along with IFN-α in late 1990s. It can be administered orally, intravenously and through inhalation (Thomas et al., 2012). A multi-center study was conducted in Brazil by Ferreira et al. (2015). It involved in the treatment of human immunodeficiency virus (HIV) patients infected from HVC sustained virological response (SVR) with PEG-IFN and RBV. It was found that with the use of this drug 11.3% suffered from adverse events mainly cytopenia and psychiatric disorders (Ferreira et al., 2015). Similarly, PEG-IFN combined with RBV was used to treat HIV and HVC patients SVR. The 40% patients experienced recurrence and 50% no serious adverse events. The use of PEG-IFN and RBV shows a good safety profile as compared to standard IFN therapy (Ke et al., 2016). Sherigar et al. (2017) studied 240 elderly HCV patients for the treatment with newer direct acting antivirals (DAAs) and with or without the use of RBV and PEG-IFN. The standard recommended therapy was continued for 12 weeks. It was concluded that the regimens were well tolerated and safely used by the elderly patients and only 6.28% patients failed to respond to the treatment indicating that the failure rate was not significant.

A follow-up study was also conducted on 4639 patients (2004-2013) treated with PEG-IFN and RBV and were examined for clinical predictors for the risk of hepatocellular carcinoma (HCC) among HCV. It was found that 233 HCC cases were reported after the follow-up of 26,163 persons. The HCV suffered patients before they developed cirrhosis showed a higher efficacy than did the treatment of those who had already developed cirrhosis. It was concluded that there is a need of regular visit for the patients with SVR because of high risk of HCC (Lee et al., 2017).

Druyts et al. (2013) have reviewed the safety and efficacy of RBV with PEG-IFN-α-2a and PEG-IFN-α-2b in adolescents and children suffered from HCV. It was noted in eight trials that 70% and 58% achieved early virological response (EVR) and SVR, respectively. The discontinuation of therapy due to adverse events and lack of response was 4% and 15%, correspondingly. It was also found that the use of RBV in combination with PEG-IFN-α-2a and PEG-IFN-α-2b is an effective and safe treatment for adolescents and children. A recent review by Maughan and Ogbuagu (2018) has been published on the treatment of HCV with PEG-IFN-α-2a alone and/or in combination with RBV. It was found that PEG-IFN-α-2a, although with prompt action under alarming adverse events, its efficacy even established in renal insufficient patients and others where DAAs have no access. It was also pointed out that knowledge of pharmacogenetics would supplement IFN-based therapy.

4.2. Telaprevir (TPV)

Telaprevir (TPV) is an oral antiviral and inhibitor of the HCV non-structural protein NS3-4A SER protease. It is used as co-therapy with PEG-IFN-α and RBV. It possesses significant efficacy against genotype-1, HCV. TPV on administration for 1 week with PEG-IFN-α and 48 weeks with RBV in HCV patients showed that TPV possessed better effect in combination with PEG-IFN-α and RBV against HCV (Perry, 2012). Kieffer et al. (2007) have found that in wild-type virus and pre-existing TPV-resistant variants of HCV alone TPV is not effective compared to TPV with PEG-IFN-α-2a in combination therapy. It was also observed that PEG-IFN-α-2a in combination with RBV possessed sensitivity against TPV-resistant virus.

The use of triple regimen therapy (i.e., PEG-IFN-α-2a or 2b, RBV and TPV or boceprevir) showed a significantly higher SVR rate than dual therapy (i.e., PEG-IFN and RBV), but dual therapy obtained a significantly higher SVR rate in patients with rapid virological response. (Coppola et al., 2014).

A survey on complexity and morbidity of TPV was conducted by Lakshmi et al. (2013). The adverse effects noted were ~ 70% (anemia), 54% (fatigue), 48% (skin rashes), 29% (anxiety and depression), 22% (anorectal discomfort), and one patient died from septic shock during the treatment. It was concluded that the treatment of HCV with TPV based regimen have higher rate of morbidity, treatment failure, relapses and serious adverse events than in published phase-3 clinical trials.

4.3. Entecavir (ETV)

Entecavir (ETV) is a potent inhibitor of HCV. ETV co-therapy with PEG-IFN increases the therapeutic response. Brouwer et al. (2015) conducted a study on patients suffering from hepatitis-B-e-antigen (HBeAg)-positive chronic hepatitis B patients with compensated liver disease on treatment with ETV. It was also observed that PEG-IFN co-therapy resulted in higher response with more viricidal effect than the ETV alone. Similarly, Xie et al. (2015) evaluated the safety and efficacy profile of ETV and IFN in hepatitis B virus (HBV). It was noted that a combination of ETV and IFN was better compared to ETV or IFN alone in safety profile. It has been shown that PEG-IFN increases the long lasting response, whereas, nucleos(t)ide analogues oddly prove off- treatment response in HBV infection. The efficacy of combination therapy of PEG-IFN-α-2a and ETV has been evaluated in HBVeAg-positive patients. It was found that sequential combination therapy has a higher rate of HBeAg seroconversion and HB-surface-antigen (HBsAg) excretion than ETV alone. The combination therapy aids to maintain baseline HBsAg and HBeAg level (Li et al., 2015).

4.4. Tenofovir (TFV)

Jindal and his co-workers (2018) used mono-therapy of PEG-IFN-α for the treatment of HBV and TFV for the add-on sequential therapy. The combination therapy did not only decrease the viral load but also was safer with an increase in serological response rate. There was a significant loss of HBeAg and HBsAg as compared to that of TFV alone. The suppression of HBsAg from serum is mainly considered as an indication of total eradication of HBV (Zhang et al., 2016). Moreover, the combination therapy of TFV and PEG-IFN-α-2a greatly reduce the level of HBsAg than that of alone PEG-IFN-α-2a or TFV (Marcellin et al., 2016). Similarly, Al-Ashgar (2017) used a combination therapy of PEG-IFN and TFV to determine the role of PEG-IFN in chronic HBV patients for HBsAg clearance. It was found that

PEG-IFN and TFV successfully achieved HBsAg clearance and seroconcersion in 4.3% patients, while mono-therapy of TFV did not achieve a decrease in HBsAg level.

4.5. Adefovir (AFV)

Korkmaz et al. (2014) conducted a study in Turkey to evaluate the efficacy of AFV PEG-IFN-α-2a. They found that virological response rate and biochemical studies of HBeAg positive and negative have shown nearly similar results.

Furthermore, the efficacy and safety profile of PEG-IFN-α and AFV co-therapy was obtained against PEG-IFN-α mono-therapy. It was found that the use of PEG-IFN-α and AFV for 48 weeks was safer and had greater efficacy than the PEG-IFN-α mono therapy, whereas, the virological and biochemical response rates were similar in both therapies (Piccolo et al., 2009). However, Zhang et al. (2016) found no statistical difference between combination therapy of PEG-IFN-α and AFV and PEG-IFN-α alone for by the HBsAg seroconversion.

4.6. Sofosbuvir (SFV)

Sofosbuvir (SFV) is a uridine nucleotide analogue (Kowdley et al., 2013). It is an HCV polymerase inhibitor that has pangenotypic antiviral activity and possess a high genetic barrier to resistance (Keating, 2014). SFV is included in triple regimen (i.e., PEG-IFN, RBV and SFV) to increase the rate of SRV in HCV genotype 1 infected patients (Dolatimehr et al., 2017). The antiviral efficacy, tolerability, and safety of the triple regimen (i.e., SFV, RBV, and PEG-IFN) has been evaluated in HCV infection (Abd-Elsalam et al., 2018). The study was conducted on 1200 chronic HCV patients, eligible for IFN therapy. The successful eradication of virus was evident by SVR.

The combination therapy of SFV, RBV, and PEG-IFN is also used to treat HCV genotype-4. Around 88%, 93% and 80% SVR was achieved in all patients, native patients and experienced patients, respectively, whereas, treatment failed in 12% patients. This combination showed fruitful results in Eygptian patients suffering from HCV genotype 4 (Abdel-Moneim et al., 2018).

The efficacy of SFV and RBV with or without PEG-IFN-α given to HCV genotype 3 patients showed that 81.7% and 84.8% patients achieved SVR in 12 weeks by dual and triple regimen, respectively (Abbas et al., 2017). Moreover, Pol et al. (2015) found that the use of SFV, PEG-IFN and RBV for 12 weeks was safe and effective in HCV-genotype-1 patients with earlier regimens of one or more DAAs plus Peg-IFN and RBV. The combination therapy of SFV and ledipasvir (LDV) with and without RBV was given once daily fixed dose. This combination was approved in many countries to treat HCV. It was found that LDV and SFV with or without RBV produced valuable treatment with good safety and efficacy profile in patients suffering from HCV-genotype-2, 3, 4, 5 or 6, pre and/or post renal or liver transplantation, blood disorders and on IFN therapy (Scott, 2018).

5. Separation and Purification

PEG is a moderately inert hydrophilic polymer and induces its effect on physicochemical properties of proteins. Thus, the size and molecular weight of fractionated proteins is different from that of the PEGylated protein. It is important to maximize the yield of PEGylated proteins because of their high cost. The purification and separation of PEG-protein is mainly carried out by consideration of its usual properties such as the molecular weight and size and electrostatic charge distribution. The ultrafiltration and chromatographic techniques like ion-exchange chromatography (IEC), size exclusion chromatography (SEC), hydrophobic interaction chromatography (HIC), reverse phase chromatography (RPC), affinity chromatography, membrane chromatography and gel *filtration chromatography* are normally used for the separation and purification of PEGylated proteins.

It is well known that PEG is a neutral polymer. The chromatographic techniques using charge development are employed to separate and purify

PEG conjugated proteins and native proteins. SEC, IEC are commonly used for the separation and purification of PEGylated proteins, whereas, HIC, MC and AC are used to a little extent for PEG-proteins separation (Fee and Van Alstine, 2006). It may be difficult to separate the native and conjugated proteins due to smaller variations in their charge affinities. It has also been observed that an increase in size of PEG polymer increases the resistance in separation and purification of proteins and conjugated proteins. It can be used for analytical scale but not efficiently possible for preparative scale separation. However, in the case of separation of the substrate and conjugated proteins, separation can be effectively achieved by SEC, membrane chromatography and ultrafiltration because these techniques are used for separation based on molecular sizes. However, the relative size difference between variants differing in PEGylation extent by one adduct reduces with *N* (number of polymer size), so that efficient SEC separation between PEGylated species differing the extent of by one PEG chain is not likely to be economic at the preparative scale for $N > 3$. It has been shown that for the larger PEG polymers ($\geq$ 20 kDa) are difficult to separated compared to smaller ($\leq$ 2kDa) ones for native and PEGylated species. For analytical scale SEC provides extensive information, whereas, the membrane chromatography separates smaller species by dialysis but complete purity and yields is not achieved (Fee and Van Alstine, 2011).

5.1. Ion-Exchange Chromatography (IEC)

The post-transitional modification needs the quality attribute of therapeutic proteins. This transition may lead to a change in pharmacokinetics profile and dramatically increase structural characteristics. The conjugation of PEG (40 kDa) with IFN-α-2b at different pH values (6.5, 7 and 8) has been observed. IEC has been used for the isolation of monoPEGylated IFN from polyPEGylated IFN and unmodified IFN. It aids in the isolation of 96% pure monoPEGylated IFN which shows a significantly better bioactivity with 330-fold increase in serum half-life (Ramon et al., 2005).

Muneeruddin et al. (2017) have used a combination of IEC with electrospray ionization mass spectrometry for the structural information of PEG-IFN-β-1a. IEC provides chromatograms with resolved peaks to represent the degradation within the protein and varying extent of PEGylation. Therefore, this combination aids in the characterization of three types of post-transitional modification within the therapeutic protein. Similarly, Ginn et al. (2016) separated PEGylated IFN-β-1b by IEC and explored its *in-vitro* bioactivity.

The PEGylated recombinant human IFN-ω by monomethoxy-PEG-succinimidyl succinate (20 kDa) was separated by IEC and gel chromatography. The average yield of purified monoPEG-recombinant human IFN-ω content exceeded to 22% (Liu et al., 2012). Furthermore, the separation and purification of PEGylated recombinant human granulocyte stimulating factor (rh-G-CSF) has been carried out by cation-exchange chromatography (CEC) to separate the PEG-rh-G-CSF from unconjugated rh-G-CSF and then by anion-exchange chromatography (AEC) to the individual PEG-rh-G-CSF (mono-, di-, and tri- PEGylated-rh-G-CSF). The results revealed that the excess free PEG and three conjugated species are purified by consecutive IEC steps. The bioactivity of the mono-, di-, and tri-PEGylated rh-G-CSF was 90%, 75%and 43%, respectively. The consecutive IECs provide high recovery and performance and are easy to scale up (Chen et al., 2005).

5.2. Size Exclusion Chromatography (SEC)

SEC is widely used for the separation of PEG conjugated protein by increase in molecular weight due to larger hydrodynamic radius than the proteins. Only by using SEC it is difficult to remove unconjugated PEG-proteins, low molecular weight impurities and positional isomer of the same molecular weight. The purification is efficiently done when the hydrodynamic radius of PEG-protein is more than 1.25-fold. SEC is used in combination with HIC and IEC for better results (Wang et al., 2000; Fee and Van Alstine, 2011). The elution of components of a PEG mixture is based on the degree of modification of the protein. The unconjugated protein is first eluted followed by mono- and poly-PEGylated protein. HIC is not suitable because of poor resolution of PEGylated proteins, hydrophobicity of proteins and binding of PEG in HIC column. However, IEC, especially CEC aids in the purification of PEG-protein poly-, mono-PEG-proteins and unconjugated proteins, respectively, in a single step due to difference in the charge and particularly, the molecular weight of PEG attached to the protein. It is better to simultaneously use CEC and AEC using cationic and anionic resins, respectively, for the elution of PEG-protein conjugates before the un-PEGylated protein (Jevsevar et al. 2010).

The viscosity of PEG solution may cause hindrance in purification by CEC. The column fouling and high back-pressure are avoided by diluting PEGylated mixtures to remove the unreacted PEG (Jevsevar et al. 2010). It aids in the determination of viscosity radii of equivalent spheres of proteins covalently bonded with PEG. The accurate viscosity radius permits calculation of the distribution coefficient, K_{av}, for PEG-proteins in SEC (Fee and Van Alstine, 2004).

The PEGylation of IFN-α-2b is carried out in preparative CEC to prepare mono-PEG-IFN-α-2b.It is used to separate LYS and HIS and CYS at pH 10 and 6.5, respectively. The purified yield of mono-PEG-IFN-α-2b is >99% by high performance size exclusion chromatography (HPSEC) (Grace et al., 2005).HPSEC is also known as Gel Permeation Chromatography (GPC) and gel filtration chromatography (GFC). The HPSEC can be attached with various detectors (e.g., light scattering, UV-vis, refractive index or photodiode array) to determine the molecular weight and radius distributions of PEG-proteins and other macromolecules. The main advantages of HPSEC over SEC are excellent reproducibility and high robustness (Liu, 2015). The mathematical derivation is used for the preparative or analytical fractionation of PEGylated mixtures. It may provide insight to several factors related to production and purification of PEGylated-protein (Mohs et al., 2014). IFN-α-2a, PEGylated by methoxy-PEG (5 kDa) is purified by SEC. The mPEG-IFN-α-2a is ~75% encapsulated in multilamellar liposomes and with 87.9% *in-vitro* antiviral activity (Vyas et al., 2006). The calibration of SEC columns is carried out by using the known molecular weight standards in terms of distribution coefficient (K_{av}) against natural logarithm of molecular weights (Kusterle et al., 2008)

5.3. Hydrophobic Interaction Chromatography (HIC)

The adsorption of bio-molecules to a weekly hydrophobic surface in high salt content is followed by descending salt gradient that aids in the purification of PEG-proteins by HIC. The molecular interaction and chemical equilibrium of rh-IFN-γ refolding is indicated by high performance HIC. It provides satisfactory results of bioactivity and mass recovery of rh-IFN-γ (Geng et al., 2004).

In HIC column, PEG shows better effect than in the dilution for the evaluation of correct refolding of consensus-IFN. The refolding yield is increased by a novel dual gradient strategy by enhancing PEG concentration and reducing guanidine-HCl concentration. Whereas, denatured consensus-IFN was absorbed and eluted from HIC column. HIC in the presence of PEG increases the specific activity 2.6-fold and an increase in soluble protein recovery up to 30% (Wang et al., 2006). Lin et al. (2001) studied the influence of relative hydrophobities of different sorbents and effect of different salts on the absorption process of proteins in HIC. It has been found that the free energy change for protein adsorption to HIC is mainly characterized by changes in enthalpy linked with the protein and sorbent dehydration and hydrophobic interaction. It has been observed that there is a difference in binding mechanisms of *n*-butyl- and phenyl-HIC sorbents. In the latter case, there is a significant change in the associated enthalpy and entropy due to the participation of π–π hydrophobic interactions. The increase in the hydrophobicity of proteins or sorbents results in more negative values of free energy, which arises mostly from the dehydration processes. However, entropic effect favors increased adsorption in HIC by an increase in non-polar protein exposure. The larger changes in entropy consist of change in interaction mechanism due to binding by adsorption.

5.4. High Performance Liquid Chromatography (HPLC)

The initial step of purification of the IFNs includes gel filtration followed by high performance liquid chromatography (HPLC). The first step involves reverse-phase HPLC on LiChrosorb RP-8 for normal partition chromatography and reverse-phase chromatography at pH 7.5 and 4.0, respectively. It has been observed that gel filtration column alone is not sufficient in purification of human IFN. A new method has been developed by using a sensitive fluorescence technique for the detection of IFN. It was possible to apply normal-phase chromatography with a diol silica column between the two RP columns for the purification of IFN-α. The actual purification of IFN was enhanced by ~ 80,000-fold (Pestka, 2007).

El Ghazalay et al. (2013) modified a RP-HPLC method for the analysis of IFN-α-2a and PEG (20 kDa) conjugated IFN-α-2a. It was found that PEG20 moiety is not detected at the wavelength of 210 nm. Equal concentrations of used internal reference standard (i.e., PEG20-IFN-α-2a) with chemical reference standard (i.e., IFN-α-2a) were found to give high correlation. Moreover, accurate yield results were measured for 10 batches of Referon Retard®.

HPLC coupled with evaporative light scattering detection (ELSC) has been used to determine modification of rh-GSF with PEG molecule. The free PEG content was determined by external standard method and double logarithm calibration curve. The method was validated for linearity, accuracy, specificity, and precision with determination of limit of detection (*LOD*) and limit of quantification (*LOQ*). It was demonstrated as a useful method for separation of an average recovery of 98.1% PEG-rh-GSF (Li et al., 2018).

5.5. Gel Electrophoresis

The characterization of PEG-IFN and free IFN can also be carried out by sodium dodecyl sulfate-capillary gel electrophoresis (SDS-CGE). The distribution profile of PEGylation by SDS-CGE was more reliable than that attained by SDS-polyacrylamide gel electrophoresis (SDS-PAGE). SDS-CGE aids in the monitoring of optimization of PEGylation such as the molar ratio of reaction. The main advantage of this technique is minimal consumption, high resolution and rapid workup (Na et al., 2004). A combination of SDS-PAGE and RP-HPLC gave 98% purity of mono-PEG-rh-IFN-ω. The molecular weight measured by SDS-PAGE was ~ 60 kDa. The antigenicity was reduced at least 64-fold. Furthermore, it was stable in serum, thermostable and acid stable with improved pharmacological properties (Liu et al., 2012).

The IFN-α-2b has been conjugated with branched PEG (40 kDa) or linear monomethoxy PEG (12 kDa). The purified monoPEG (40)-IFN-α-2b shows transparent, colorless bands against zinc imidazolate-generated white gel background with in 20 min. It has been confirmed by IEC based on positional isomer separation profile and enzyme-linked immunosorbent response assay (ELISA). It is useful for the rapid determination or small scale PEGylation of IFN (Hardy et al., 2008).

The IFN-β-1b has been purified completely by Blue Sepharose column chromatography. The added CYS residue did not interfere with proper folding and disulfide bonding. The non-reducing and reducing SDS-PAGE has been used to recover most of CYS analogs of IFN-β-1b monomers that co-migrated with IFN-β-1b (Lee et al., 2013).

5.6. Mass Spectrometry (MS)

Mass spectrometry (MS) is used to confirm the conjugation site of PEG with proteins. It also aids in the determination of molecular weight of PEGylated proteins. MS not only provides the description of qualitative estimation but also the quantitative evaluation of free or PEGylated drugs in the blood, and to explore the PEGylated protein structure (Hutanu and Darie, 2014). In a previous study, rh-IFN-α-2a has been PEGylated at the *N-terminus* with 5, 10 and 20 kDa mPEG aldehyde by reductive alkylation. Where, the matrix-assisted laser desorption ionization -time-of-flight

(MALDI-TOF) MS has confirmed the conjugation of PEG molecule only to the *N-terminus* (Lee et al., 2007).

Chen et al. (2005) have used MALDI-TOF for the determination of molecular weights of mono-, di- and tri-PEGylated rh-GSF having molecular weights of 23.8, 28.6 and 33.8 kDa, respectively. The *in-vitro* bio-activity was 90%, 75% and 43%, for mono-, di- and tri-PEG-rh-GSF than unconjugated rh-GSF.

The conjugation of PEG-protein mainly occurs at *N-terminus*, sulfhydryl, carbohydrate and on amino acid like GLU, CYS, ASP, SER, TRY, ARG, LYS, HIS, THR. The MS is an accurate and robust analytical technique for protein characterization, post-translational modifications of proteins and their interaction studies. MALDI-TOF MS provides accurate average weight and extent of PEGylation. It can easily analyze different types of PEGs (i.e., mono, linear, branched, hetrofunctional), monodisperse and/or polydisperse) attached to protein ranging in size ~ 5-50 kDa. The size of proteins is changed due to covalent bonding and increase in structural hetrogeneity. However, the concept of molecular mass is not taken into consideration because different glycoforms of PEGylated protein have nearly identical distribution of ionic charges as a function of *m/z*. However, MALDI-MS is significant in determination molecular weight, heterogeneity, and especially total amount and distribution of PEG and site of conjugation. Furthermore, electrospray ionization (ESI) MS is preferred over MALDI because of automated reduced sampling time and workflow (Hutanu and Darie, 2014). MALDI-MS showed ~95% monoPEGylation of IFN-α-2b by 12 kDa PEG. It has been found that ~ 50% monoPEGylation of IFN-α-2b on the HIS^{34} residue gives better antiviral activity (Grace et al., 2001). The molecular weight of PEG-rh-IFN-ω as 43790 Da has been observed by MALDI-TOF MS. The pharmacological properties of PEG-IFN-ω are significantly increased with around 6-fold decrease in antigenicity (Liu et al., 2012).

6. STABILITY OF IFN AND PEG-IFN

The use of therapeutic proteins may encounter many physical problems, like unfolding and aggregation and chemical degradation such as oxidation, proteolysis, disulfide scrambling, and deamidation etc. It should also be ensured that the efficacy and safety of therapeutic proteins must retain the physicochemical properties from manufacturing until the end use. Apart from environmental factors like light, temperature, oxygen, and shipping and handling agitation, while packaging materials also play a part in the integrity of proteins (Patel et al., 2011). The chemical and physical degradation of proteins and peptides may lead to immunogenic hazards. The glycosylated proteins after post-transitional modifications, like phosphorylation undergo to significant degradation. They are also sensitive to a small change in osmolarity and pH of solutions. Even freeze-drying is quite sensitive and can cause degradation in the presence of different stabilizers. The stability and stabilization of protein drugs is still a major concerned for the desired therapeutic effect (Lipiainen et al., 2015). The packaging material and chemical modification are used to stabilize protein drugs. Thus PEGylation increases the stability against thermal, pH and enzymatic degradation (Patel et al., 2011).

The chemical bond also affects the stability of proteins, like that in two therapeutic IFNs, e.g., PEG-IFN-α-2b has a urethane bond which is unstable and leads to hydrolysis. This may results in serum release of IFN-α-2b after administration of PEG-IFN-α-2b. It is reconstituted before administration and discarded without storage. However, PEG-IFN-α-2a has an amide bond that is chemically stable and intact (Foster, 2004; Singal, 2010).

A molecular dynamic stimulation (MDS) study has been carried out to establish the involvement of all-atoms in conjugation between different PEG polymers and insulin. It was noted that PEG conjugates around insulin by hydrophobic interaction and forms parallel hydrogen bonds with the surrounding water molecule and thus enhances water stability. Moreover, it leads to prolonged half-life and prevents from proteolysis, immunologic reactions, kidney excretion, and enzymatic degradation. Thus, the optimal chain length maximizes insulin potency (Yang et al., 2011). The biophysical and structural characterization of PEGylated proteins is determined by NMR, circular dichroism (CD), fluorimetry and dynamic scanning calorimetry (DSC). It has been proved that a compact, flexible, and mobile random coil is attached to the protein. PEG-rh-IFN-α-2a has more effective hydrodynamic volume, higher thermostablity and decreased tendency of aggregation. Thus, it is an important step toward the understanding of the molecular level of the binding of PEG-IFN-α-2a and its positional isomers to its cellular receptors (Dhalluin et al., 2005).

The PEGylated of IFN-α using mPEG (2 or 5 kDa) is encapsulated in biodegradable, PLGA, microspheres. The conjugation was characterized by SDS-PAGE, SE-HPLC, MALDI-TOF-MS. It stabilized on exposure to dicholoromethane. The release of native IFN was found to be 16.3±0.4%, whereas, PEG_{2k}-IFN and PEG_{5k}-IFN were released to the extent of 72.5±2.1 and 56.8±2.5%, respectively. Therefore, long-acting IFNs can be stabilized by using the combination of PEGylation and encapsulation (Diwan and Park, 2003).

Tsuge et al. (2017) developed a novel site specific PEG-IFN-β (TRK-560) and *in-vitro* and *in-vivo* evaluation on HBV replication. There was a significant antiviral activity observed by cDNA microarray analysis and enzyme-linked immunoassay (ELISA). The activity is enhanced by TFV and ETV. The novel TRK-560 has shown better efficacy against PEG-IFN-α-2a with loss of HBsAg and HBV-covalently closed circular(ccc)-DNA eradication (Tsuge et al., 2017).

Ginn et al. (2013) used site specific PEGylation to integrate the protein folding process. The conjugation of IFN-β with PEG stabilizes the protein structure during refolding (Ginn, 2013).

The production of copper catalyzed reactive oxygen species may degrade amino acid residues. The use of copper-catalyzed azide-alkyne cycloaddition (CuAAC) aids to produce site-specific modified IFN-β-1b by PEGylation. It was observed that the degradation by oxidation was increased with the reaction time, while a decrease in copper (Cu) concentration reduced the PEGylation rate as well as oxidative degradation. CYS as a reducing agent has been used for the PEGylation of other proteins (Nairn et al., 2015).

The physical and chemical stability of PEG-IFN-α-2b has been tested under different stress conditions like pH, temperature, mechanical agitation, and repeated freeze-thaw. A validated SE-HPLC method can be used to determine the stability. After incubation at 37°C, pH 4.0-1.0, for a period of 4 weeks resulted in more degradation compared to that at the same temperature and pH 2.0-8.0 for 24 h. This degradation was not noted for 30 months at 2-8°C. Thus, the stability of this protein maintained at refrigeration (Abed et al., 2014).

The IFN-α possesses a shorter half-life in serum. The hybridization of IFN-α with human immunoglobin (Fc) may increase the half-life. A compression study of stability, tissue distribution, molecular imaging and pharmacokinetic of ^{131}I-IFN-α and ^{131}I-IFN-α-Fc has been carried out in rats. It was shown that ^{131}I-IFN-α-Fc has better stability than ^{131}I-IFN-α in plasma, blood, kidney, stomach, and liver. The area under curve (AUC) of ^{131}I-IFN-α-Fc was 10-fold higher and its degradation was slower. The hybridized cytokines can be used for long-term treatment (Hsu et al., 2015).

The IFN-β-1b has been PEGylated with 12, 20, 30, 40 kDa amine selective PEG polymers at the *N-terminus* or randomly on the 11 LYS to increase the serum half-life and bioavailability. The antiviral and antiproliferation bioassays have shown the stability of PEG-IFN-β-1b as study by several techniques such as MALDI-TOF, circular dichroism, capillary electrophoresis, flow cyctometry, RP-HPLC and SE-HPLC. The well tolerated immunogenicity of the drug proved its efficacy, safety and stability profile (Basu et al., 2006).

Ruiz with his co-workers (2005) conducted a study to evaluate the effect of packaging material on the stability of rh-IFN-α-2b. The compatibility of IFN with packaging materials, chlorobutyl stoppers and type-I borosilicate glass, additives like tween-80 and EDTA and buffer species such as phosphate, citrate and citophosphate was studied by ELISA and RP-HPLC. The effect of heat degradation was observed in 1 ml of rh-IFN- α-2b, whereas, the instability was not detected in 0.5 ml of the protein. The use of chlorobutyl stopper increased the degradation rate from 6.74 ± 0.38 to $46.34 \pm 3.11 \times 10^3$ day^{-1}. Citrate and citrophosphate buffers decreased the instability ~ 1.2 and 1.1-fold, respectively. The additives decreased the degradation rate constant range from 2.01 ± 0.14 to 25.51 ± 3.57 x 10^3 day^{-1}. It was noted that the chlorobutyl stopper increases the degradation. The study was further extended by Ruiz et al. (2006) who determined the aggregation of rh-IFN-α-2b in solution. There was an increase in the rate of aggregation of rh-IFN-α-2b in sodium phosphate buffer from 1.55 to 1.8 × 10^3 day^{-1} was determined by SDS-PAGE. Moreover, it was also noted that the presence of divalent metal ions like Zn^{2+} and Cu^{2+} increases the degradation kinetics of rh-IFN-α-2b due to metal-catalyzing reactions.

The irreversible aggregation of therapeutic proteins is a major problem. The prevention of physical degradation of rh-IFN-α-2b is matter of great concern. The effect of heat, pH, and buffer concentrations on the stability of rh-IFN-α-2b has been studied by SE-HPLC, dynamic light scattering and SDS-PAGE methods. The rh-IFN-α-2b is aggregated to form higher molecular weight (measured at OD_{350}). It has also been observed that at 50°C the formation of aggregates is increased up to 240 h. The aggregation is also increased at pH 6.0 than that of pH 5.0 and pH 7.0 with an increase in buffer concentration. Thus, a change in solution formulation would change the rh-IFN-α-2b stability (Salmannejad et al., 2014). Similarly, it was also found that the stability of rh-IFN-α-2b is influenced by a change in the concentration of therapeutic protein itself at 37°C. The higher the content of protein, the higher would be the stability. However, the purity of rh-IFN-α-2b is better determined at low concentration (1-10 MIU/ml) than at the higher concentration (50-100 MIU/ml) by SDS-PAGE and RP-HPLC. The accelerated stability analysis has shown good chemical stability for 30 months in the absence of albumin-free formulation at the recommended storage temperature i.e., 5± 3°C (Ruiz et al., 2006).

Malek-Sabet et al. (2013) investigated the effect of oxygen on the oxidative degradation of rh-IFN-γ. The rh-IFN-γ dimmers and deaminated forms were analyzed by HPLC, SE-HPLC and SDS-PAGE and it was found that their quantity is increased ~ 11.5%. Furthermore, after 9 months the antiviral activity against vesicular stomatitis virus is also decreased to 41%. Thus, oxygen tension produces significant effect on the stability of rh-IFN-γ (Malek-Sabet et al., 2013). The physical stability of rh-IFN-γ in guanidine and urea solution has been determined by fluorescence spectrometry. The denaturation of rh-IFN-γ occurred more in guanidine than in urea at pH 7.4. The effect of electrostatic interaction did not play a significant role in the destabilization of rh-IFN-γ. It was pointed out that protonation of ASP leads to conformational changes which has an impact on the stability of rh-IFN-γ (Christova et al., 2003).

The effect of benzyl alcohol on unfolding and aggregation of IFN-α-2a has been studied. It is used as a preservative in marketed IFN-α-2a preparations. It was demonstrated by isothermal aggregation and temperature scanning that by an increase in the concentration of benzyl alcohol the apparent aggregation temperature of IFN-α-2a is linearly decreased. This has been confirmed by NMR chemical shifts and hydrogen exchange rates. Thus, the preservatives used should be at an optimal concentration or changed to achieve stable IFN-α-2a (Bis et al., 2015).

The glycosylation of rh-IFN-β-1a by site-directed mutagenesis results in a new molecular identity. It does not affect the ligand-receptor binding and provides enhanced stability with higher half-life and reduced aggregation (Song et al., 2014).

The use of hydrogen peroxide (H_2O_2) to mediate oxidation of rh-IFN-β-1a and increase the immunogenicity has been studied. The oxidized IFN has shown enhancement in oligomer content and uncovered protein content. The treatment of IFN with oxidizing agents leads to aggregation and causes immunologic reactions (van Beers et al., 2011). Diress et al. (2010) investigated the aggregation, denaturation, and reduction of IFN-α-2a and IFN-α-2b by SE-HPLC with fluorescence detection to easily measure even the minor changes. The aggregates were formed around 650 kDa with dimmers, reduced and denatured variants. Moreover, the changes may decrease 87% of the biological activity of IFN-α-2b. The list of marketed PEGylated approved FDA and European Union are presented in Table 2.

Table 2. Pegylated marketed products (CPW, 2017)

Generic Name	Brand Name	PEG size (kDa)	Clinical use	Route of administration	Year of Approval
PEG-ademase bovine	Adagen	5	Immunodefficiency	i.m.	1990
PEG-aspargase	Oncaspar	5	Leukemia	i.v. or i.m.	1994
PEG-interferon-alfa-2b	PegIntron	12	Hepatitis C	s.c.	2001
PEG-interferon-alfa-2a	Pegasys	40	Hepatitis B	s.c.	2002
PEG-visomant	Somavert	5	Acromegaly	s.c.	2002
PEG-filgrastim	Neulasta	20	Neutropenia	s.c.	2002
PEG-aptanib sodium	Macugen	40	Wet form of age-related macular degradation	intravitreal	2004
Continuous erythropoietin receptor activator mPEG-epoetin	Mircera	30	Anemia	i.v. & 0s.c.	2007
PEGylated Fab fragment of the humanized anti-tumor necrosis factor (TNF)-alpha monoclonal antibody certolizumab pegol	Cimzia	40	Crohn's disease, psoriatic arthiritis	i.v. & s.c.	2008
PEG-loticase	Krystexxa	10	Chronic gout	i.v.	2010
PEG-inesatide	Omontys	40	Anemia	i.v./s.c.	2012 (Recall in February 2013)
PEG-interferon-beta-1	Plegridy	20	Multiple sclerosis	s.c.	2014
Pegylated naloxol	Movantik	3.35	Opioid-induced constipation	oral	2014

7. Future Prospects on PEG-IFNs

There is a considerable scope in future for the development of PEG-IFN formulations for novel applications, such as for renal cell carcinoma (kidney cancer), chronic myeloid leukaemia (myeloproliferative neoplasm) and metastatic melanoma and other infections in following area.

1. Entrapment in liposomes for targeted delivery
2. Encapsulation in biodegradable polymers in combination with DAAs
3. Formulation in the form of inhalers
4. Formulation with nanoparticles for nanocarrier drug delivery
5. Formulated as short acting antiviral vaccines.

Conclusion

There is a breakthrough in the eradication of hepatic viral infections by PEG-IFN. These agents have shown remarkable effect in combination with RBV in the treatment of liver infection. The invention of DAAs has revolutionized the treatment of these infections with better efficacy and safety profile. The DAAs have initiated research in the development and invention of new antiviral agents. It was reported that >90% SVR has been achieved with newly available combinational regimens. Clinical and pharmacological research is needed to improve the effectiveness and safety profile of these drugs. Although PEG-IFN is a relatively high cost therapy, but it has been noted in many studies that DAAs in combination with PEG-IFN produce prolonged effect with an increase in efficacy and better safety in liver infections. Moreover, PEG-IFNs could be used particularly in different cancer treatments. There is still an ample space in research and development of novel formulations involving IFNs and PEG-IFNs.

REFERENCES

Abbas, Z., Saad, M., Nadeem, R., Jawed, F., Abbas, M., et al. (2017). Sofosbuvir and ribavirin with or without PEGylated interferon for hepatitis C genotype 3: a real world experience, *Hepatttis Monthly*, *17*(4), e12297.

Abdel-Aziz, M.S. & Janson, J.C. (2012). Hydrolytic and chromatographic studies on the PEGylation of dextranase from *Penicillium* sp. *Journal of Genetic Engineering and Biotechnology*, *10*(1), 145-150.

Abdel-Moneim, A., Abood, A., Abdel-Gabaar, M., Zanaty, M.I. & Ramadan, M. (2018). Effectiveness of sofosbuvir/pegylated-interferon plus ribavirin in treatment of hepatitis C virus genotype 4 patients. *Clinical and Experimental Hepatology*, 2018, 4, 3, 191–196.

Abd-Elsalam, S., Badawi. R., Elnawasany, S., Yousef, M., Mansour, L., Hawash, N., et al. (2018). Sofosbuvir, pegylated interferon and ribavirin in treatment of an Egyptian cohort with hepatitis C virus infection in real life clinical practice. *Infectious Disorders Drug Targets*, doi: 10.2174/1871526518666180912121835.

Abed, H.S., Al-Ghobashy, M.A., Fathalla, F.A. & Salem, M.Y. (2014). Assessment of the degradation pattern and extent of PEG interferon α-2b using a stability-indicating SE-HPLC assay. *Chromatographia*, *77*, 1661.

Ahad, M.A., Alim, M.A. & Ekram, A.R.M.S. (2004). Interferon to PEG-interferon: a review. *Transactional Analysis Journal*, *17*(2), 113-116.

Akbarzadehlaleh, P., Mirzaei, M., Mashahdi-Keshtiban, M., Shamsasenjan, K. & Heydari, H. (2016). PEGylated human serum albumin: review of pegylation, purification and characterization methods. *Advanced Pharmaceutical Bulletin*, *6*(3), 309-317.

Al Ashgar, H., Peedikayil, M. C., Al Quaiz, M., Al Sohaibani, F., Al Fadda, A., Khan, M. Q., et al. (2017). HBsAg clearance in chronic hepatitis B patients with add-on pegylated interferon alfa-2a to ongoing tenofovir treatment: a randomized controlled study. *Saudi Journal of Gastroenterology*, *23*(3), 190-198.

Awwad, S., Ginn, C. & Brocchini, S. (2018). The case for protein PEGylation. In: *Engineering of Biomaterials for Drug Delivery Systems Beyond Polyethylene Glycol*, Parambath AK (Ed.),Woodhead Publishing Series in Biomaterials, Elsevier Publishers, Amsterdam, Netherlands, pp. 27-49.

Baeshen, N.A., Baeshen, M.N., Sheikh, A., Bora, R.S., Ahmed, M.M.M., Ramadan, H.A.I., Saini, K.S. & Redwan, E.M. (2014). Cell factories for insulin production. *Microbial Cell Factories.*, *13*, 141.

Banerjee, S.S., Aher, N., Patil, R. & Khandare, J. (2012). Poly(ethylene glycol)-prodrug conjugates: concept, design, and applications. *Journal of Drug Delivery*, 2012, 103973.

Basu, A., Yang, K., Wang, M., Liu, S., Chintala, R., Palm, T., et al. (2006).Structure−function engineering of interferon-β-1b for improving stability, solubility, potency, immunogenicity, and pharmacokinetic properties by site-selective mono-PEGylation, *Bioconjugate Chemistry*, *17* (3), 618-630.

BC Cancer Agency (BCCA) (2016), *Cancer Drug Manual* https://www.accessdata.fda.gov/drugsatfda_docs/label/2011/103964s5204lbl.pdf (Accessed on 10th November, 2018).

Behi, J., Hassiki, R., Said, N.B., Bouhaouala-Zahar, B. & Benkhoud, M.L. (2018). Optimization of PEGylation reaction time and molar ratio of rhG-CSF toward increasing bioactive potency of monoPEGylated protein. *International Journal of Biological Macromolecules*, *109*, 888-895.

Bis, R.L., Singh, S.M., Cabello-Villegas, J. & Mallela, K.M.G. (2015). Role of benzyl alcohol in the unfolding and aggregation of interferon α-2a. *Journal of Pharmaceutical Sciences*, *104*(2), 407-415.

Brouwer, W.P., Xie, Q., Sonneveld, M.J., Zhang, N., Zhang, Q., Tabak, F., Streinu-Cercel, A., Wang, J.Y., et al. (2015). Adding PEGylated interferon to entecavir for hepatitis B e antigen-positive chronic hepatitis B: A multicenter randomized trial (ARES study). *Hepatology*, *61*(5), 1512-1522.

Bruno, R., Sacchi, P., Cima, S., Maiocchi, L., Novati, S., Filice, G. & Fagiuoli, S. (2012). Comparison of peginterferon pharmacokinetic and pharmacodynamic profiles. *Journal of Viral Hepatitis*, *19*(1), 33-36.

Burnham, N.L. (1994). Polymers for delivering peptides and proteins. *American Journal of Hospital Pharmacy*, *51*(2), 210-218; quiz 228-9.

Chavez, M.D., Schrimsher, J.L. & Morar, A.S. (2006). PEGylation of proteins: a structural approach. *BioPharm International*, *19*(4), 34-49.

Chen, T., Yun, Q., Bi, J.X., Ma, G.H., Su, Z.G., (2005). Separation and purification of PEGylated rhG-CSF by two-step ion-exchange chromatography. *Sheng Wu Gong Cheng Xue Bao.*, 2005 Mar, *21*(2), 284-8.

Chiu, K., Agoubi, L.L., Lee, I., Limpar, M. T., Lowe, J. W. Jr. & Goh, S.L. (2010). Effects of polymer molecular weight on the size, activity, and stability of PEG-functionalized trypsin. *Biomacromolecules*, *11*(12), 3688-3692.

Cho, Y.W., Park, J.H., Park, J.S. & Park, K. (2007). PEGylation: camouflage of proteins, cells, and nanoparticles against recognition by the body's defense mechanism. In: *Handbook of Pharmaceutical Biotechnology*, Gad SC (Ed.), John Wiley & Sons, Inc. New Jersey, USA, pp. 443-462.

Christova, P., Todorova, K., Timtcheva, I., Nacheva, G., Karshikoff, A. & Nikolov, P. (2003). Fluorescence studies on denaturation and stability of recombinant human interferon-gamma. *Z Naturforsch C*, *58*(3-4), 288-294.

Coppola, N., Pisaturo, M., Sagnelli, C., Sagnelli, E. &Angelillo, I.F. (2014). Peg-interferon plus ribavirin with or without boceprevir or telaprevir for HCV genotype 1: a meta-analysis on the role of response predictors. *PLoS One*, *9*(4), e94542.

Creative PEG Works (CPW). (2017). *Marketed PEGylated Drugs Approved by FDA and in Europe.* https://creativepegworks.wordpress.com/2017/03/07/marketed-pegylated-drugs-approved-by-fda-and-in-europe/ (Accessed on 10th November, 2018).

D'souza, A.A. & Shegokar, R. (2016). Polyethylene glycol (PEG): a versatile polymer for pharmaceutical applications. *Expert Opinion on Drug Delivery*, *13*(9), 1257-1275.

Da, Silva., Freitas, D., Mero, A. & Pasut, G. (2013). Chemical and enzymatic site specific PEGylation of hgh. *Bioconjugation Chemistry*, *24*, 456-463.

Dalsgaard, T.K. & Larsen, L.B. (2009). Effect of photo-oxidation of major milk proteins on protein structure and hydrolysis by chymosin. *International Dairy Journal*, *19*(6–7), 362-371.

Damodaran, V.B. & Fee, C. (2010). Protein PEGylation: an overview of chemistry and process considerations. *European Pharmaceutical Review*, *15*, 18-26.

Dhalluin, C., Ross, A., Leuthold, L.A., Foser, S., Gsell, B., Muller, F. & Senn, H. (2005). Structural and biophysical characterization of the 40 kDa PEG-interferon-alpha2a and its individual positional isomers. *Bioconjugate Chemistry*, *16*(3), 504-517.

Di Bisceglie, A.M., Ghalib, R.H., Hamzeh, F.M. &Rustgi, V.K. (2007). Early virologic response after peginterferon alpha-2a plus ribavirin or peginterferon alpha-2b plus ribavirin treatment in patients with chronic hepatitis C. *Journal of Viral Hepatitis*, *14*(10), 721-729.

Diress, A., Lorbetskie, B., Larocque, L., Li, X., Alteen, M., Isbrucker, R. & Girard, M. (2010). Study of aggregation, denaturation and reduction of interferon alpha-2 products by size-exclusion high-performance liquid chromatography with fluorescence detection and biological assays. *Journal of Chromatography A*, *1217*(19), 3297-3306.

Diwan, M. & Park, T.G. (2003). Stabilization of recombinant interferon-alpha by pegylation for encapsulation in PLGA microspheres. *International Journal of Pharmaceutics*, *252*(1-2), 111-122.

Dolatimehr, F., Karimi-Sari, H., Rezaee-Zavareh, M.S., Alavian, S.M., Behnava, B., Gholami-Fesharaki, M. & Sharafi, H. (2017). Combination of sofosbuvir, pegylated-interferon and ribavirin for treatment of hepatitis C virus genotype 1 infection: a systematic review and meta-analysis. *DARU Journal of Pharmaceutical Sciences*, *25*(1), 11.

Dolence, E.K., Hu, C., Tsang, R., Sanders, C.G. & Osaki, S. (1997). *Electrophilic polyethylene oxides for the modification of polysaccharides, polypeptides (proteins) and surfaces*, US Patent, 5, 650, 234.

Dozier, J.K. & Distefano, M.D. (2015). Site-specific PEGylation of therapeutic proteins. *International Journal of Molecular Sciences*, *16*(10), 25831-25864.

Druyts, E., Thorlund, K., Wu, P., Kanters, S., Yaya, S., Cooper, C.L. & Mills, E.J. (2013). Efficacy and safety of pegylated interferon alfa-2a or alfa-2b plus ribavirin for the treatment of chronic hepatitis C in children and adolescents: a systematic review and meta-analysis. *Clinical Infectious Diseases*, *56*(7), 961-967.

El Ghazaly, M., Meager, A., Zikry, H., Ebaed, M., Shaker, S., Mueller, F. &Rohde, J. (2013). Quantification of a PEGylated interferon-alpha2a product by a customised and validated reverse phase-high performance liquid chromatography method. *Journal of Pharmaceutical and Biomedical Analysis*, *84*, 48-52.

Farhadi, T. & Hashemian, S.M.R. (2018). Computer-aided design of amino acid-based therapeutics: a review. *Drug, Design, Development and Therapy.*, *12*, 1239–1254.

Fee, C.J. & Van Alstine, J.M. (2004). Prediction of the viscosity radius and the size exclusion chromatography behavior of PEGylated proteins. *Bioconjugation Chemistry*, *15*(6), 1304-1313.

Fee, C.J. & Van Alstine, J.M. (2006). PEG-proteins: reaction engineering and separation issues. *Chemical Engineering Science*, *61*(3), 924-993.

Fee, C.J. & Van Alstine, J.M. (2011). Purification of pegylated proteins. *Methods of Biochemical Analysis*, *54*, 339-362.

Ferreira, P.R.A., da Silva, M.H., Brandao-Melo, C.E., Rezende, R.E., et al. (2015). The clinical effectiveness of PEGylated interferon and ribavirin for the treatment of chronic hepatitis C in HIV-infected patients in Brazil: a multicentric study. *Brazilian Journal of Infectious Diseases*, *19*(1), 15-22.

Food Drug Administration (FDA). (2003). https://www.accessdata.fda.gov/drugsatfda_docs/label/2001/pegsche080701lb.htm (Accessed on 10th November, 2018).

Foster, G.R. (2004). Review article: PEGylated interferons: chemical and clinical differences. *Alimentary Pharmacology and Therapeutics*, *20*(8), 825-830.

Geng, X., Bai, Q., Zhang, Y., Li, X. & Wu, D. (2004). Refolding and purification of interferon-gamma in industry by hydrophobic interaction chromatography. *Journal of Biotechnology*, *113*(1-3), 137-49.

Gibbert, K., Schlaak, J.F., Yang, D. & Dittmer, U. (2013). IFN-α subtypes: distinct biological activities in anti-viral therapy. *British Journal of Pharmacology*, *168*(5), 1048-1058.

Ginn, C., Choi, J.W. & Brocchini, S. (2016). Disulfide-bridging PEGylation during refolding for the more efficient production of modified proteins. *Biotechnology Journal*, *11*(8), 1088-1099.

Ginn, C.L. (2013). *Protein PEGylation on Protein Folding*. Doctoral thesis, UCL (University College London).

Goldberg, M. & Gomez-Orellana, I. (2003). Challenges for the oral delivery of macromolecules. *Nature Reviews Drug Discovery*, *2*(4), 289-295.

Grace, M., Youngster, S., Gitlin, G., Sydor, W., Xie, L., Westreich, L., Jacobs, S., Brassard, D., Bausch, J. & Bordens, R. (2001). Structural and biologic characterization of PEGylated recombinant IFN-alpha2b. *Journal of Interferon & Cytokine Research*, *21*(12), 1103-1115.

Grace, M.J., Lee, S., Bradshaw, S., Chapman, J., Spond, J., Cox, S., Delorenzo, M., Brassard, D., Wylie, D., Cannon-Carlson, S., Cullen, C., Indelicato, S., Voloch, M. & Bordens, R. (2005). Site of PEGylation and polyethylene glycol molecule size attenuate interferon-alpha antiviral and antiproliferative activities through the JAK/STAT signaling pathway. *The Journal of Biological Chemistry*, *280*(8), 6327-6336.

Hamley, I.W. (2014). PEG–peptide conjugates. *Biomacromolecules*, *15*(5), 1543-1559.

Hardy, E., Ramon, J., Saez, V., Baez, R., Tejeda, Y. & Ruiz, A. (2008). Detection of PEGylated proteins in polyacrylamide gels by reverse staining with zinc and imidazole salts. *Electrophoresis*, *29*(11), 2363-2371.

Hou, W., Wang, X., Ye, L., Zhou, L., Yang, Z.Q., Riedel, E. & Ho, W. Z. (2009). Lambda interferon inhibits human immunodeficiency virus type 1 infection of macrophages. *Journal of Virology*, *83*(8), 3834-3842.

Hsu, W.C., Chien, Y.C., Chang, C.H., Yuan, T.T., Lee, T.W. & Hwang, J.J. (2015). Characteristic Comparison Between 131I-Interferon-α and

131I-Interferon-α-Immunoglobulin-Fc Hybrid Protein in Rats Using Molecular Imaging. *In Vivo*, *29*(4), 445-52.

Hutanu, D. & Darie, C.C. (2014). Trends in Characterization of PEGylated Proteins by Mass Spectrometry. *Modern Chemistry and Application*, *2*, 128.

Jang, H.J., Shin, C.Y. & Kim, K.B. (2015). Safety evaluation of polyethylene glycol (PEG) compounds for cosmetic use. *Toxicological Research*, *31*(2), 105-136.

Jevsevar, S., Kunstelj, M. & Porekar, V.G. (2010). PEGylation of therapeutic proteins. *Biotechnology Journal*, *5*(1), 113-128.

Jindal, A., Vyas, A.K., Kumar, D., Kumar, G., Sharma, M.K. & Sarin, S.K. (2018). Higher efficacy of PEGylated interferon-α2b add-on therapy in hepatitis B envelope antigen-positive chronic hepatitis B patients on tenofovir monotherapy. *Hepatology Research*, *48*(6), 451-458.

Joklik, W.K. (1977). The mechanism of action of interferon. *Annals of The New York Academy of Science*, *284*(1), 711-716.

Kamaly, N., Yameen, B., Wu, J. & Farokhzad, O.C. (2016). Controlled-release polymers and polymeric nanoparticles: mechanisms of controlling drug release. *Chemical Reviews*, *116*(4), 2602–2663.

Ke, Y.C., Li, L.H., Hu, F.Y., He, Y., Lan, Y., Chen, X.J., Tang, X.P. & Cai, W.P. (2016). Clinical efficacy of pegylated interferon in patients co-infected with HIV and HCV who failed standard interferon therapy. *Zhonghua Gan Zang Bing Za Zhi*, *24*(3), 181-185.

Keating, G.M. (2014). Sofosbuvir: a review of its use in patients with chronic hepatitis C. *Drugs*, *74*(10), 1127-1146.

Kieffer, T. L., Sarrazin, C., Miller, J. S., Welker, M. W., Forestier, N., et al. (2007). Telaprevir and PEGylated interferon-alpha-2a inhibit wild-type and resistant genotype 1 hepatitis C virus replication in patients. *Hepatology*, *46*(3), 631-639.

Klaus, W., Gsell, B., Labhardt, A.M., Wipf, B.& Senn, H. (1997). The three-dimensional high resolution structure of human interferon alpha-2a determined by heteronuclear NMR spectroscopy in solution. *Journal of Molecular Biology, 274*(4):661-675.

Kontsek, P. (1994). Human type I interferons: structure and function. *Acta Virologica*, *38*(6), 345-360.

Korkmaz, P., Usluer, G., Ozgunes, I., Kartal, E.D., Erben, N. & Alpat, S.N. (2014). Comparison of adefovir dipivoxil and pegylated interferon alpha-2a treatment in chronic hepatitis B patients. *Northern Clinics of Istanbul*, *1*(1), 26-32.

Kouchakzadeh, H., Shojaosadati, S. A., Maghsoudi, A. & Vasheghani Farahani, E. (2010). Optimization of PEGylation conditions for BSA nanoparticles using response surface methodology. *AAPS PharmSciTech*, *11*(3), 1206-1211.

Kowdley, K.V., Lawitz, E., Crespo, I., Hassanein, T., Davis, M.N., DeMicco, M., et al. (2013). Sofosbuvir with PEGylated interferon alfa-2a and ribavirin for treatment-naive patients with hepatitis C genotype-1 infection (ATOMIC): an open-label, randomised, multicentre phase 2 trial. *Lancet*, *381*(9883), 2100-2107.

Krause, C.D., Digioia, G., Izotova, L.S., Xie, J., Kim, Y., Schwartz, B.J., Mirochnitchenko, O.V.& Pestka, S. (2013). Ligand-independent interaction of the type I interferon receptor complex is necessary to observe its biological activity. *Cytokine*. *64*(1):286-297.

Krishnamurthy, R. & Manning, M.C. (2002). The stability factor: importance in formulation development. *Current Pharmaceutical Biotechnology*, *3*(4), 361-371.

Kusterle, M., Jevsevar, S. & Porekar, V.G. (2008). Size of Pegylated Protein Conjugates Studied by Various Methods. *Acta Chimica Slovenica*, *55*(3), 594-601.

Lawrence, P.B. & Price, J.L. (2016). How PEGylation influences protein conformational stability. *Current Opinion Chemical Biology*, *34*, 88-94.

Le Page, C., Genin, P., Baines, M.G. & Hiscott, J. (2000). Interferon activation and innate immunity. *Reviews of Immunogenetics*, *2*(3), 374-386.

Lee, B.K., Kwon, J.S., Kim, H. J., Yamamoto, S. & Lee, K.E. (2007).Solid-phase PEGylation of recombinant interferon α-2a for site-specific modification: process performance, characterization, and *in vitro* bioactivity, *Bioconjugate Chemistry*, *18*(6), 1728-1734.

Lee, J.I., Eisenberg, S.P., Rosendahl, M.S., Chlipala, E. A., Brown, J. D., et al. (2013). Site-specific PEGylation enhances the pharmacokinetic properties and antitumor activity of interferon beta-1b. *Journal of Interferon Cytokine Research*, *33*(12), 769-777.

Lee, M.H., Huang, C.F., Lai, H.C., Lin, C.Y., et al. (2017). Clinical efficacy and post-treatment seromarkers associated with the risk of hepatocellular carcinoma among chronic hepatitis C patients. *Scientific Report*, *7*(1), 3718.

Lee, S. & McNemar, C. (1999). *Substantially pure histidine-linked protein polymer conjugates*, US Patent, *5*, 985,263.

Li, G.J., Yu, Y.Q., Chen, S.L., Fan, P., Shao, L.Y., Chen, J.Z., et al. (2015). Sequential combination therapy with PEGgylated interferon leads to loss of hepatitis B surface antigen and hepatitis B e antigen (HBeAg) seroconversion in HBeAg-positive chronic hepatitis B patients receiving long-term entecavir treatment. *Antimicrobial Agents and Chemotherapy*, *59*(7), 4121-4128.

Li, Y.H., Shi, X.C., Li, X., Han, C.M. & Rao, C.M. (2018). Development and validation of a method for determination of average modification extent of PEGylated recombinant human granulocyte colony-stimulating factor by high performance liquid chromatography coupled with evaporative light scattering detection. *Chinese Journal of Biologicals*, *31*(7), 758-762.

Lin, F.Y., Chen, W.Y. & Hearn, M.T. (2001). Microcalorimetric studies on the interaction mechanism between proteins and hydrophobic solid surfaces in hydrophobic interaction chromatography: effects of salts, hydrophobicity of the sorbent, and structure of the protein. *Analytical Chemistry*, *73*(16), 3875-3883.

Lipiainen, T., Sarkhel, P.S., Yrjonen, T., Vuorela, H., Urtti, A. & Juppo, A. (2015). Formulation and stability of cytokine therapeutics. *Journal of Pharmaceutical Sciences*, *104*(2), 307-326.

Liu, H., Cheng, Y.G., Pan, H.C., Xu, B., Peng, L., Yang, H.T. & Guo, W. (2012). Preparation and characterization of mono PEGylated recombinant human interferon omega. *Yao Xue Xue Bao*, *47*(3), 393-398.

Liu, X.M. (2015). The Importance and Applications of High Performance Size Exclusion Chromatography in Biopharmaceutical and Medical Dcvice Industries. *Symbiosis Open Access Journal of Chromatographic Science, 1*(1), 1.

Malek-Sabet, N., Masoumian, M.R., Zeinali, M., Khalilzadeh, R. & Mousaabadi, J.M. (2013). Production, purification, and chemical stability of recombinant human interferon-γ in low oxygen tension condition: a formulation approach. *Preparative Biochemistry and Biotechnology*, *43*(6), 586-600.

Manneberg, M., Lahm, H.W. & Fountoulakis, M. (1995). Oxidation of cysteine and methionine residues during acid hydrolysis of proteins in the presence of sodium azide. *Analytical Biochemistry*, *224*(1), 122-127.

Marcellin, P., Ahn, S.H., Ma, X., Caruntu, F.A., Tak, W.Y., et al. (2016). Combination of tenofovir disoproxil fumarate and peginterferon α-2a increases loss of hepatitis b surface antigen in patients with chronic hepatitis B. *Gastroenterology*, *150*(1), 134-144.e10.

Marshall, S.A., Lazar, G.A., Chirino, A.J. & Desjarlais, J.R. (2003). Rational design and engineering of therapeutic proteins. *Drug Discovery Today*, *8*(5), 212-221.

Maughan, A. & Ogbuagu, O. (2018). Pegylated interferon alpha 2a for the treatment of hepatitis C virus infection. *Expert OpiniononDrug MetabolismandToxicology*, *14*(2), 219-227.

Meng, F., Manjula, B.N., Smith, P.K. & Acharya, S.A. (2008). PEGylation of human serum albumin: reaction of PEG-phenyl-isothiocyanate with protein. *Bioconjugation Chemistry*, *19*(7), 1352-1360.

Metz, D.H. (1975). The mechanism of action of interferon. *Cell.*, *6*(4), 429-439.

Meyer, J.D., Ho, B. & Manning, M.C. (2002). Effects of conformation on the chemical stability of pharmaceutically relevant polypeptides. *Pharmaceutical Biotechnology*, *13*, 85-107.

Miron, T. & Wilchek, M. (1993). A simplified method for the preparation of succinimidyl carbonate polyethylene glycol for coupling to proteins. *Bioconjugate Chemistry*, *4*, 568-569.

Mishra, P., Nayak, B. & Dey, R.K. (2016). PEGylation in anti-cancer therapy: an overview. *Asian Journal of Pharmaceutical Sciences*, *11*(3), 337-348.

Mohs, A., Ambrogelly, A., Yang, X., Haverick, M., Cheung, J.K., Narasimhan, C. & Shameem, M. (2014). Effect of Pegylation on Self-Association of IFN-α2b. *Molecular Pharmaceutics*, *11*(1), 158-163.

Morgenstern, J., Baumann, P., Brunner, C. & Hubbuch, J. (2017). Effect of PEG molecular weight and PEGylation degree on the physical stability of PEGylated lysozyme. *International Journal of Pharmaceutics*, *519*(1-2), 408-417.

Muneeruddin, K., Bobst, C.E., Frenkel, R., Houde, D., Turyan, I., Sosic, Z. & Kaltashov, I.A. (2017). Characterization of a PEGylated protein therapeutic by ion exchange chromatography with on-line detection by native ESI MS and MS/MS. *The Analyst*, *142*(2), 336-344.

Na, D.H., Park, E.J., Youn, Y.S., Moon, B.W., Jo, Y.W., Lee, S.H., et al. (2004). Sodium dodecyl sulfate-capillary gel electrophoresis of polyethylene glycolylated interferon alpha. *Electrophoresis*, *25*(3), 476-479.

Naik, A.S., Srivatsa, K., Mane, K., Kodali, P. & Adhikary, L. (2017). Site directed PEGylation: Role of pH and temperature, to increase the productivity of insulin oligomer conjugate. *Journal of Pharmaceutical and Biological Sciences*, *5*(4), 155-160.

Nairn, N.W., Bariola, P.A., Graddis, T.J., Van Brunt, M.P., Wang, A., Li, G. & Grabstein, K. (2015). Cysteine as a Monothiol Reducing Agent to Prevent Copper-Mediated Oxidation of Interferon Beta During PEGylation by CuAAC. *Bioconjugate Chemistry*, *26*(10), 2070-2075.

Noureddin, M. & Ghany, M.G. (2010). Pharmacokinetics and pharmacodynamics of PEGinterferon and ribavirin: implications for clinical efficacy in the treatment of chronic hepatitis C. *Gastroenterology Clinics of North America*, *39*(3), 649-658.

Patel, J., Kothari, R., Tunga, R., Ritter, N.M. & Tunga, B.S. (2011). Stability considerations for biopharmaceuticals, Part 1 overview of protein and peptide degradation pathways. *Bioprocess International*, 20–31.

Paul, F., Sandra, Pellegrini S. & Gilles, Uze G. (2015). IFNA2: The prototypic human alpha interferon. *Gene*, 567(2): 132–137.

Pedder, S.C. (2003). Pegylation of interferon alfa: structural and pharmacokinetic properties. *Seminars in Liver Disease*, *23*(Suppl 1), 19-22.

Pestka, S. (2007). The interferons: 50 years after their discovery, there is much more to learn. *The Journal of Biological Chemistry*, *282*(28), 20047-20051.

Piccolo, P., Lenci, I., Demelia, L., Bandiera, F., Piras, M.R., Antonucci, G., et al. (2009). A randomized controlled trial of PEGylated interferon-alpha2a plus adefovir dipivoxil for hepatitis B e antigen-negative chronic hepatitis B. *Antiviral Therapy*, 2009, *14*(8), 1165-1174.

Product Information Leaflet (PIL). (1997). https://www.accessdata.fda.gov/drugsatfda_docs/label/1997/ifnasch110697-lab.pdf (Accessed on 11th Nov, 2018).

Pol, S., Sulkowski, M.S., Hassanein, T., Gane, E.J., Liu, L., Mo, H., et al. (2015). Sofosbuvir plus PEGylated interferon and ribavirin in patients with genotype 1 hepatitis C virus in whom previous therapy with direct-acting antivirals has failed. *Hepatology*, *62*(1), 129-134.

Potakamuri, L.N., Thuluvath, P.L., Kantsevoy, S. & Maheshwari, A. (2013). The experience with telaprevir based HCV therapy in community practice does not mirror the clinical trials data. *Gastroentrology*, *144*(5, Suppl.1), S-985.

Priyanka, R. Muralidharan. (2014). Interferons and interferon therapy. *Journal of Pharmaceutical Science and Research*, *6*(12), 400-403.

Quadt-Akabayov, S.R., Chill, J.H., Levy, R., Kessler, N. & Anglister, J. (2006). Determination of the human type I interferon receptor binding site on human interferon-α2 by cross saturation and an NMR-based model of the complex. *Protein Science, 15*(11), 2656-2668.

Radwanski, E., Perentesis, G., Jacobs, S., Oden, E., Affrime, M., Symchowicz, S. & Zampaglione, N. (1987). Pharmacokinetics of interferon alpha-2b in healthy volunteers. *Journal of Clinical Pharmacology*, *27*(5), 432-435.

Rajender, R.K., Modi, M.W. & Pedder, S. Use of PEGinterferon alfa-2a (40 KD) (Pegasys) for the treatment of hepatitis C. *Advanced Drug Delivery Reviews*, *54*(4), 571-586.

Ramon, J., Saez, V., Baez, R., Aldana, R. & Hardy, E. (2005). PEGylated interferon-alpha2b: a branched 40K polyethylene glycol derivative. *Pharmaceutical Research*, *22*(8), 1374-1386.

Roberts, M.J., Bentley, M.D. & Harris, J.M. (2002). Chemistry for peptide and protein PEGylation. *Advanced Drug Delivery Reviews*, *54*(4), 459-476.

Rodriguez, J.L., Valenzuela, C., Marin, N., et al. (2000). Comparative pharmacokinetics and pharmacodynamics of two recombinant human interferon alpha-2b formulations administered intra muscularly in healthy male volunteers. *Biotecnología Aplicada*, *17*(3), 166-170.

Ruiz, L., Aroche, K. & Reyes, N. (2006).Aggregation of recombinant human interferon alpha 2b in solution: Technical note. *AAPS Pharm Sci Tech*, *7*(4), E118-E122.

Ruiz, L., Reyes, N., Aroche, K., Báez, R., Aldana, R. &Hardy, E. (2006). Some factors affecting the stability of interferon alpha 2b in solution. *Biologicals*, *34*(1), 15-19.

Ruiz, L., Reyes, N., Aroche, K., Tolosa, V., Simanca, V., Rodriguez, T. & Hardy, E. (2005). Influence of packaging material on the liquid stability of interferon-α2b. *Journal of Pharmacy and Pharmaceutical Sciences*, *8*(2), 207-216.

Salmannejad, F., Nafissi-Varcheh, N., Shafaati, A. & Aboofazeli, R. (2014). Study on the effect of solution conditions on heat induced-aggregation of human alpha interferon. *Iranian Journal of Pharmaceutical Research*, *13*(Suppl), 27-34.

Samuel, C.E. (2001). Antiviral actions of interferons. *Clinical Microbiology Reviews*, *14*(4), 778-809.

Samuel, C.E. (2001). Antiviral actions of interferons. *Clinical Microbiology Reviews*, *14*(4), 778-809.

Schuller, J., Czejka, M.J., Schernthaner, G., Wirth, M., Bosse, C., Jager, W. & Micksche, M. (1992). Pharmacokinetic aspects of interferon alfa-2b after intrahepatic or intraperitoneal administration. *Seminars in Oncology*, *19* (2 Suppl 3), 98-104.

Scott, L.J. (2018). Ledipasvir/sofosbuvir: a review in chronic hepatitis C. *Drugs*, *78*(2), 245-256.

Sherigar, J.M., Gayam, V., Khan, A., Mukhtar, O., Arefiev, Y., Khalid, M., Siddiqui, I., Rangaraju, A.M., Budhathoki, N., Mansour, M., Guss, D. & Mohanty, S.R. (2017). Clinical efficacy and tolerability of direct-acting antivirals in elderly patients with chronic hepatitis C. *European Journal of Gastroenterology and Hepatology*, *29*(7), 767-776.

Singal, A.K. & Mon, H. (2010). Type of PEGylated interferon matters: another milestone in the treatment of hepatitis C virus infection. *Summer*, *10*(3), 165-167.

Song, K., Yoon, I.S., Kim, N.A., Kim, D.H., Lee, J. & Lee, H.J. (2014). Glycoengineering of interferon-β 1a improves its biophysical and pharmacokinetic properties. *PLoS One*, *9*(5), e96967.

Stetson, D.B. & Medzhitov, R. (2006). Antiviral defense: interferons and beyond. *Journal of Experimental Medicine*, *203*(8), 1837-1841.

Stuart-Harris, R.C., Lauchlan, R. & Day, R. (1992). The clinical application of the interferons: a review. NSW Therapeutic Assessment Group. *Medical Journal of Australia*, *156*(12), 869-872.

Suo, X., Lu, X., Hu, T., Ma, G. & Su, Z. (2009). A solid-phase adsorption method for PEGylation of human serum albumin and staphylokinase: preparation, purification and biochemical characterization. *Biotechnology Letters*, *31*(8), 1191-1196.

Thomas, C., Moraga, I., Levin, D., Krutzik, P.O., Podoplelova, Y., Trejo, A., Lee, C., Yarden, G., Vleck, S.E., Glenn, J.S., Nolan, G.P., Piehler, J., Schreiber, G. & Garcia, K.C. (2011). Structural linkage between ligand discrimination and receptor activation by type I interferons. *Cell*, *146*(4), 621-32.

Thomas, E., Ghany, M.G. & Liang, T.J. (2012). The application and mechanism of action of ribavirin in therapy of hepatitis C. *Antiviral Chemistry and Chemotherapy*, *23*, 1-12.

Top, A., Roberts, C.J. & Kiick, K.L. (2011). Conformational and aggregation properties of a PEGylated alanine-rich polypeptide. *Biomacromolecules*, *12*(6), 2184-2192.

Tsuge,, M., Uchida, T., Hiraga, N., Kan, H., Makokha, G. N. & Abe-Chayama, H., et al. Development of a novel site-specific PEGylated interferon beta for antiviral therapy of chronic hepatitis B virus. *Antimicrobial Agents and Chemotherapy*, *61*(6), e00183-17.

van Beers, M.M., Sauerborn, M., Gilli, F., Brinks, V., Schellekens, H. & Jiskoot, W. (2011). Oxidized and aggregated recombinant human interferon beta is immunogenic in human interferon beta transgenic mice. *Pharmaceutical Research*, *28*(10), 2393-2402.

van Vught, R., Pieters, R.J. & Breukink, E. (2014). Site-specific functionalization of proteins and their applications to therapeutic antibodies. *Computational and Structural Biotechnology Journal*, *9*, e201402001.

Vandermeulen, G.W.M., Tziatzios, C. & Klok, H.A. (2003). Reversible self-organization of poly(ethylene glycol)-based hybrid block copolymers mediated by a de novo four-stranded α-helical coiled coil motif. *Macromolecules*, *36*(11), 4107-4114.

Veronese, F.M. & Mero, A. (2008). The impact of PEGylation on biological therapies. *BioDrugs*, *22*(5), 315-329.

Vyas, S.P., Rawat, M., Rawat, A., Mahor, S. & Gupta, P. N. (2006). Pegylated protein encapsulated multivesicular liposomes: a novel approach for sustained release of interferon alpha. *Drug Development and Industrial Pharmacy*, *32*(6), 699-707.

Wang, F., Liu, Y., Li, J., Ma, G. & Su, Z. (2006). On-column refolding of consensus interferon at high concentration with guanidine-hydrochloride and polyethylene glycol gradients. *The Journal of Chromatography A.*, *1115*(1-2), 72-80.

Wang, Y.S., Youngster, S., Bausch, J., Zhang, R., McNemar, C. &Wyss, D.F. (2000). Identification of the major positional isomer of PEGylated interferon alpha-2b. *Biochemistry*, *39*(35), 10634-10640.

Wills, R.J. (1990). Clinical pharmacokinetics of interferons. *Clinical Pharmacokinetics*, *19*(5), 390-399.

Winger, M., de Vries, A.H. & van Gunsteren, W.F. (2009). Force-field dependence of the conformational properties of α,ω-dimethoxy-polyethylene glycol, *Molecular Physics*, *107*(13), 1313-1321.

Wu, J., Zhao, C., Lin, W., Hu, R., Wang, Q., Chen, H., Li, L., Chen, J. & Zheng, J. (2014). Binding characteristics between polyethylene glycol (PEG) and proteins in aqueous solution. *Journal of Material Chemistry B*, *2*, 2983-2992.

Xie, Q.L., Zhu, Y., Wu, L.H., Fu, L.L. & Xiang, Y. (2015). The efficacy and safety of entecavir and interferon combination therapy for chronic hepatitis B virus infection: a meta-analysis. *PLoS One*, *10*(7), e0132219.

Yang, C., Lu, D. & Liu, Z. (2011). How PEGylation enhances the stability and potency of insulin: a molecular dynamics simulation. *Biochemistry*, *50*(13), 2585-2593.

Zalipsky, S., Seltzer, R. & Menon-Rudolph, S. (1992). Evaluation of a new reagent for covalent attachment of polyethylene glycol to proteins. *Biotechnology and Applied Biochemistry*, *15*, 100-114.

Zeuzem, S., Welsch, C. & Herrmann, E. (2003). Pharmacokinetics of PEGinterferons. *Seminars in Liver Disease*, *23* (Suppl 1), 23-28.

Zhang, Y., Chen, B., Wang, L., Chi, J., Song, S., Liu, M. & Zhao, Z. (2016). HBsAg seroclearance or seroconversion induced by PEG-interferon alpha and lamivudine or adefovir combination therapy in chronic hepatitis B treatment: a meta-analysis and systematic review. *Spanish Journal of Gastroenterology (REED)*, *108*(5), 263-70.

Zheng, L., Li, M.P., Gou, Z.P., Wang, Y., Xu, N., Cai, Y.M. & Luo, H. (2015). A pharmacokinetic and pharmacodynamic comparison of a novel PEGylated recombinant consensus interferon-α variant with peginterferon-α-2a in healthy subjects. *British Journal of Clinical Pharmacology*, *79*(4), 650-659.

In: Advances in Chemistry Research
Editor: James C. Taylor
ISBN: 978-1-53615-828-1

Chapter 3

ANTHRACENE BASED LIGANDS: SYNTHESIS AND APPLICATIONS

Pooja Dubey, Alpesh Kumar Sharma and Ajai Kumar Singh*

Department of Chemistry, Indian Institute of Technology, Delhi, New Delhi, India

ABSTRACT

Anthracene has been functionalized with various donor groups such as azomethine, chalcogen, phosphorus, etc. to generate ligands of varying dentate character. The ligands are reported to react with metal precursors of different types, forming complexes of different geometries established by X-ray diffraction on their single crystals (if grown successfully). The fluorescence properties of anthracene-based ligands are suitable to design sensors for cations as well as anions. Schiff bases synthesized with anthracene skeleton have shown cytotoxic and fungicidal activity. Organometallic compounds designed with functionalized anthracene and other co-ligands are reported. MOF's are generated with anthracene-based ligands. Metal complexes of chalcogenated anthracenes efficiently

* Corresponding Author's E-mail: aksingh@chemistry.iitd.ac.in.

catalyze *N*-alkylation of amines and transfer hydrogenation reactions. The catalytic activation of C-C coupling reactions with Pd complexes of anthracene based ligands is reported to be promising. Some of the complexes of anthracene-based ligands are reported to have potential for light emitting devices.

Keywords: ligand, fluorescence, complex, metal, sensor, catalysis, biological activity

INTRODUCTION

Anthracene is a solid polycyclic aromatic hydrocarbon consisting of three fused benzene rings. With naphthalene and phenanthrene, it is among the major components of the neutral fractions of high boiling coal-tar distillates. It is a low molecular weight polycyclic aromatic hydrocarbon (PAH) that is often used as a model PAH in bioremediation study, due to its structure that is found in high molecular weight PAH's. Anthracenes undergo cycloaddition on irradiation with UV light [1]. Thus they can be an essential component of photo-switchable compounds. Owing to rigidity of the skeleton, the anthracene unit is often used when the defined distance between substituents is intended [2].

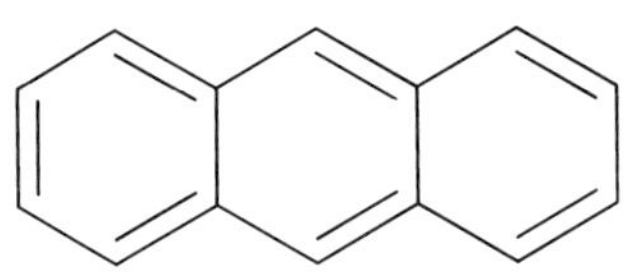

Anthracene

The presence of anthracene skeleton in ligands may exhibit three obvious effects: (i) its larger conjugated π-system results in π–π stacking and C–H–π interactions which may play a role in the formation of stable complexes by such ligands, (ii) steric effect of the bulky anthracene ring may facilitate the creation of coordinatively unsaturated species important in catalysis (iii) the presence of bulky anthracene rings may provide suitable conditions for the formation of the intra- and/or inter-molecular interactions. Anthracene derivatives are evidenced as excellent fluorophores and therefore anthracene based ligands have potential for the developments of fluorescence sensors.

SCHIFF BASES

A good number of anthracene-based Schiff bases and metal complexes of several such species are reported. The fluorescence of anthracene-based Schiff bases and their metal complexes are responsible for their potential for sensors. The various types of biological properties are associated with these species. The designing of anthracene based Schiff bases, their complexation with metallic species and diverse applications of both are focused in this section. 9-Anthrylidene-*p*-toluidine (**1**) and 9-anthrylidene-furfurylamine (**2**) were prepared by condensation of 9-anthracenecarboxaldehyde with *p*-toluidine and furfurylamine, respectively. The addition of dimethyl-phosphite and diethylphosphite to these Schiff bases according to the Scheme 1 [3] resulted in α-aminophosphonic acid diesters, bearing anthracene ring viz. [*N*-methyl(dimethoxyphosphonyl)-1-(9-anthryl)]-*p*-toluidine (**3**), [*N*-methyl(diethoxyphosphonyl)-1-(9-anthryl)]-*p*-toluidine (**4**) and [*N*-methyl(diethoxyphosphonyl)-1-(9-anthryl)]furfurylamine (**5**). Among these **2** and **4** were found most potent cytotoxic agents towards colon carcinoma cell line HT-29. The latter compound also exhibited antiproliferative activity to HBL-100, MDA-MB-231 and 647-V cells. The **2** and its synthetic precursor 9-anthrylidene-*p*-toluidine were found to be cytotoxic to HBL-100 and HT-29 tumor cell lines, respectively.

Anthracene-based receptors containing Schiff base functionality (**6**) were synthesized (Scheme 2) in one step and fully characterized. The optical properties of the two receptors towards metal ions investigated by UV–visible and fluorescence spectroscopy have revealed that both the receptors show high selectivity towards Cu(II) over other 17 metal ions in CH_3CN–H_2O (9/1,v/v) [4].

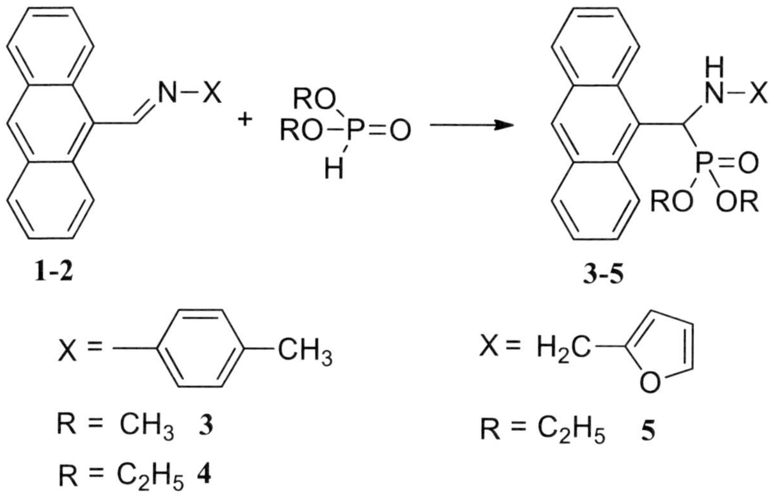

Scheme 1.

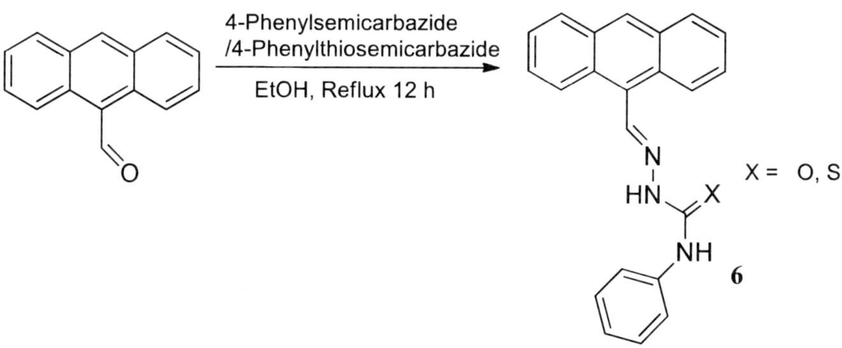

Scheme 2.

The praseodymium, erbium and ytterbium complexes of Schiff bases prepared by condensation of 3,4-diaminopyridine with 9-anthraldehyde are reported. They show antibacterial activity against Bacillus subtilis, Staphylococcus aureus, Escherichia coli and Pseudomonas aeruginosa. The Pr complexes showed better immune behavior against all the pathogens than the other two metal complexes [5]. An anthracene-based Schiff base, **7** (Scheme 3) was synthesized and characterized using FT-IR, ^{1}H and $^{13}C\{^{1}H\}$ NMR and HR-MS [6]. The propensity of the Schiff base to sense effectively

Fe^{3+} and CN^- ions was studied using UV–vis, fluorescence spectroscopy and density functional theory techniques. The colorimetric and fluorescence sensing of cyanide ion with the Schiff base in acetonitrile medium is promising [6]. The Schiff base ligand–CN^- system has been successfully used for sequential recognition of Fe^{3+} ion among various cations. The limit of detection for cyanide was found to be 8.11×10^{-6} M. The fluorescence turn-on response of the Schiff base with cyanide ion was successfully applied to bio-imaging of cell line RAW264.7. The relay recognition of Fe^{3+} was applied to construct IMP and INH logic gates. 2-((*E*)-(((E)-Anthracen-9-ylmethylene)hydrazono)methyl)-5-(diethylamino)phenol (**8**) has been synthesized and used to detect hypochlorite (Scheme 4) [7].

The **8** has shown selective turn-off fluorescence for ClO^- through oxidation reaction between imine bond and oxidizer ClO^-. It also could detect ClO^- over various competitive anions including reactive oxygen species. The photophysical properties and sensing mechanism of **8** toward ClO^- have been suggested by UV-vis and fluorescent spectroscopy, NMR titration, ESI-MS and DFT calculations.

Scheme 3.

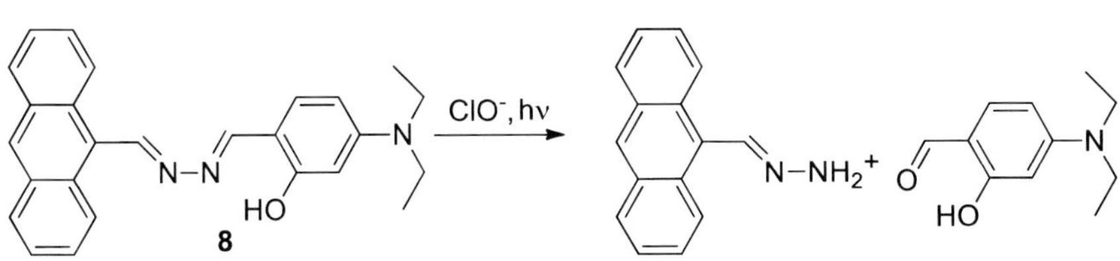

Scheme 4.

The Schiff base (9-$C_{14}H_9$)CH=NCH_2CH_2N=CH(9-$C_{14}H_9$) (**9**) is reported to react with [Pt_2Me_4(μ-SMe_2)$_2$] giving platinum(II) complex, [$PtMe_2${(9-$C_{14}H_9$)CH=NCH_2CH_2N=CH(9-$C_{14}H_9$)}] (**10**), which contains a chelating dinitrogen ligand. The **10** was reacted with MeI to give the related six coordinate platinum(IV) compound, [$PtMe_3$I{(9-$C_{14}H_9$)CH=NCH_2 CH_2N=CH(9-$C_{14}H_9$)}] (**11**), by oxidative addition. On refluxing **10** in toluene, mono-metalated complex, [PtMe{(9-$C_{14}H_9$)CH=NCH_2 CH_2N=CH (9-$C_{14}H_8$)}] (**12**), with a six-membered, cyclometalated, anthryl ring was formed.

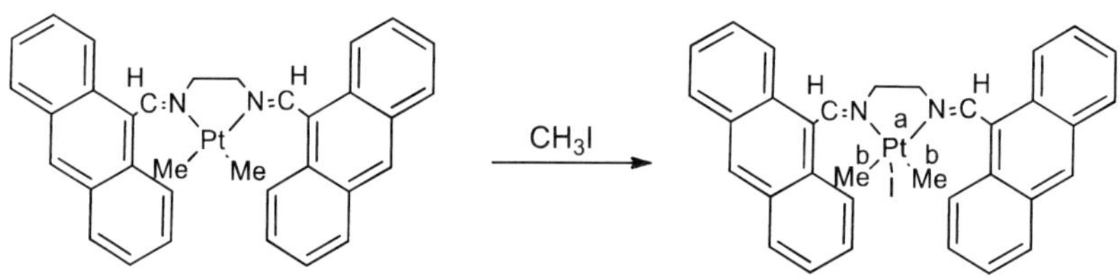

10 **11**

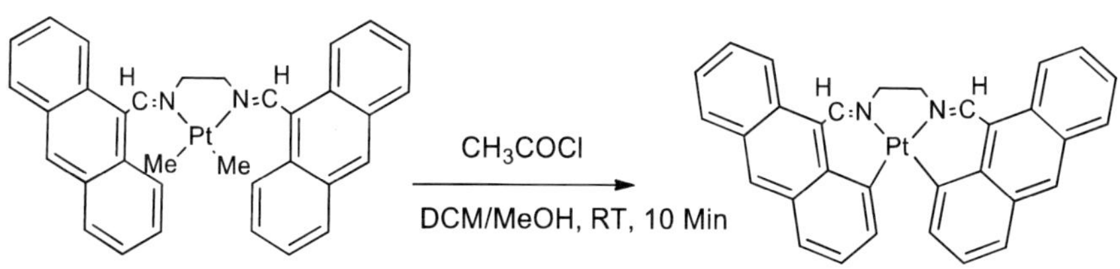

10 **13**

When the compound **10** was reacted with acetyl chloride, dicyclometalated compound, [Pt{(9-$C_{14}H_8$)-CH=NCH_2CH_2N=CH(9-$C_{14}H_8$)}] (**13**), with two six-membered cyclometallated anthryl rings, was formed. Similarly, *cis*-[$PtCl_2$(dmso)$_2$] was reacted with the ligand **9** to afford the platinum(II) compound, [$PtCl_2${(9-$C_{14}H_9$)CH=NCH_2CH_2N=CH(9-$C_{14}H_9$)}] (**14**). This compound, when left in solution at room temperature, gave the mono-metalated compound, [PtCl{(9-$C_{14}H_9$)CH=NCH_2CH_2N=CH-(9-$C_{14}H_8$)}](**15**), but under no condition, the dimetalated species was obtained. The single crystal structures of **11**, **14**, and **15**, determined by X-ray diffraction, are reported.

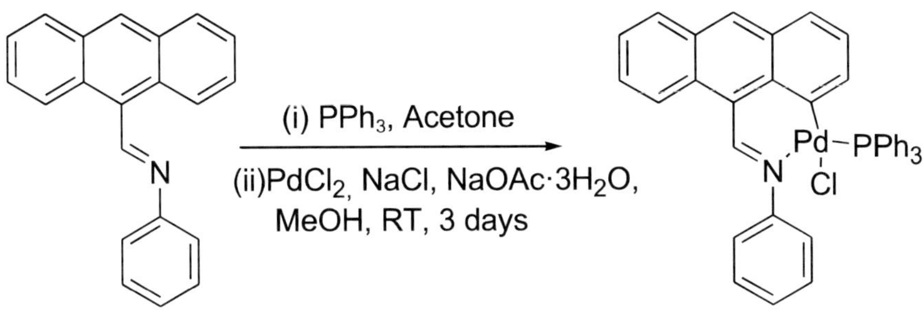

Scheme 5.

The $[Pt_2Me_4(\mu\text{-}SMe_2)_2]$ on reaction with $Me_2NCH_2CH_2NCHAr$ (Ar = 9-anthracenyl) in acetone at room temperature produced $[PtMe_2\{9\text{-}(Me_2NCH_2CH_2NCH)C_{14}H_9\}]$ (**16**) having imine coordinated as a bidentate [N, N] ligand. The refluxing of toluene solution of **16** resulted in a cyclometallated compound $[PtMe\{9\text{-}(Me_2NCH_2CH_2NCH)C_{14}H_8\}]$ (**17**). The reactions of **17** with acetyl chloride and methyl iodide produced, respectively, compounds $[PtCl\{9\text{-}(Me_2NCH_2CH_2NCH)C_{14}H_8\}]$ (**18**) and $[PtMe_2I\{9\text{-}(Me_2NCH_2CH_2NCH)C_{14}H_8\}]$ (**19**) [8]. The cyclometallation of anthracen-9-ylmethylene-phenylamine with palladium(II) acetate and with the system $PdCl_2$, NaCl and $NaOAc\cdot 3H_2O$ occurred via the C1-H bond activation of the anthracen-9-yl group, affording entry to a new type of endocyclic six-membered cyclopalladated organic imine that contains a metallated aromatic carbon atom (Scheme 5).

The centrosymmetric di-*μ*-chlorido-bis({9-[(2,6-diisopropylphenyl) iminomethyl]anthracen-1-yl}-palladium(II) (**20**) was obtained by a C—H bond-activation reaction of a anthracene based Schiff base with Li_2PdCl_4 in methanol. The single crystal structure of **20** is reported [9].

The condensation of the sodium salt of 3,5-dialkyl sulfanilic acid and 2,6-dialkyl aniline with anthracene-9-carbaldehyde in anhydrous methanol have yielded Schiff base ligands. The cyclopalladated complexes of these ligands (Scheme 6) show high catalytic efficiency [10] for Suzuki coupling of ArBr with phenylboronic acid in aqueous solvents under mild conditions. The **21** is reported to give high conversions in five Suzuki reaction cycles carried out in neat water without additives. The catalytic process for the Suzuki coupling is proved by TEM analysis to proceed on Pd(0) NPs.

However, the surfactant-protected palladium NPs exhibit lower activity and poorer recyclability for the coupling reaction than those generated *in situ* without additives.

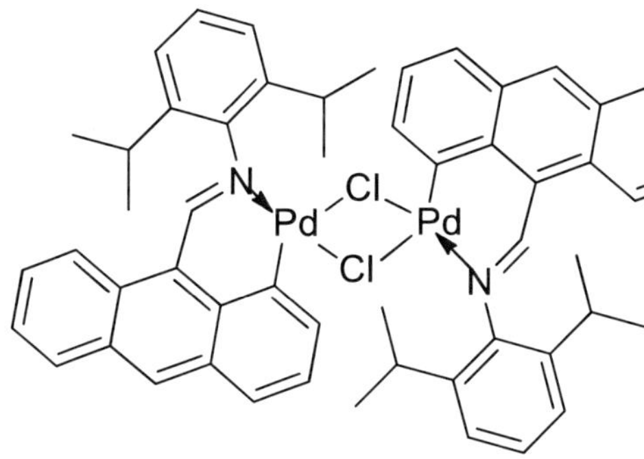

20

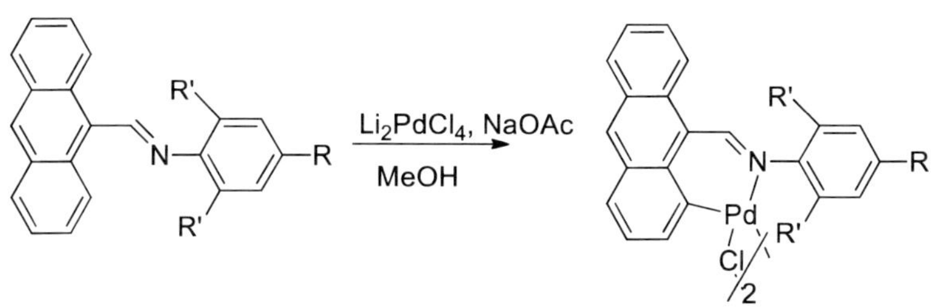

21

Scheme 6.

Copper(II) complexes of an anthracene-based ligand, N,N'-bis[anthracene-2-ylmethyl]-N,N'-bis[carboxymethyl]-1,3-diaminopropan-2-ol (**22**) have been reported (Scheme 7). The reaction of stoichiometric amounts of $CuSO_4{\cdot}5H_2O$ and **22** in the presence of NaOH at ambient temperature in methanol-dichloromethane, afforded a dinuclear copper complex, [Cu_2(**22**-3H)(MeO)$(MeOH)_2$] (**23**). The reaction of stoichiometric amounts of $Cu(OAc)_2{\cdot}H_2O$ and the **22** in the presence of NaOH at ambient temperature in methanol-dichloromethane yielded, a novel tetranuclear copper complex, [Cu_4(**22**-3H)$_2$(l-OAc)$_2$(MeOH)$_2$]$\cdot$2$CH_2Cl_2CH_3OH$ (**24**). The X-ray single structure analyses show that the complex **23** consists a distorted square planar copper center along with a distorted square pyramidal copper center with Cu···Cu separation of 3.589(4) Å, whereas in **24** four distorted square pyramidal copper centers with average Cu···Cu separation of 3.358 (6) Å are present. The molecular architecture of **23** is built from the self-assembly of two neutral [Cu_2(**23**-3H)(l-OAc)] units, which are exclusively bridged by two carboxylate functionalities of the ligand. The variable-temperature (2–300 K) magnetic susceptibility measurements on powdered microcrystalline samples show that the paramagnetic copper centers in **24** are antiferromagnetically coupled to one another. The reactions of equimolar amounts of Li_2PdCl_4, 9-anthraldehyde 4-R-benzoylhydrazones (H_2**L**n; n = 1 and 2 for R = H and NMe_2, respectively) and $CH_3COONa{\cdot}3H_2O$ in methanol produce the cyclopalladated complexes [11] of formula [Pd(H**L**n)Cl] [**25** (R = H) and **26** (R = NMe_2)].

The reactions of **25** and **26** with 2 mol equivalents of PPh_3 in acetone resulted in the complexes of formula [Pd(**L**n)(PPh_3)] [**27** (R = H) and **28** (R = NMe_2)]. The molecular structures of **25**-**28** have been established by X-ray crystallography. In **25** and **26**, the C,N,O-donor (H**L**n)$^-$ and the chloride form a CNOCl square-plane around the metal centre, whereas (**L**n)$^{2-}$ and the PPh_3 assemble a CNOP square-plane around the metal centre in **27** and **28.**

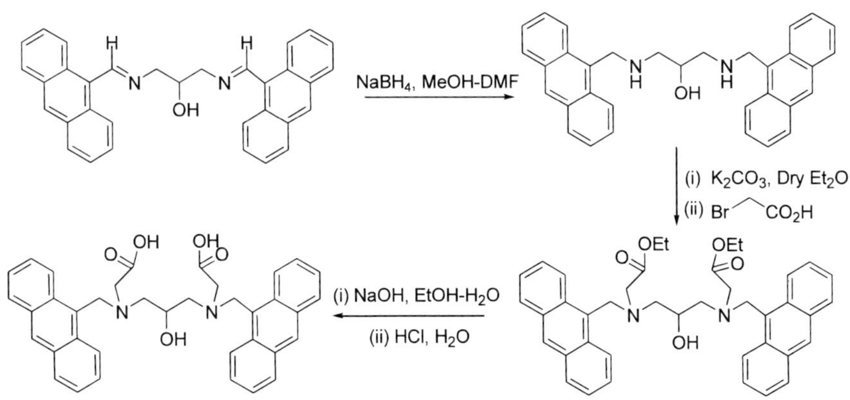

22

Scheme 7.

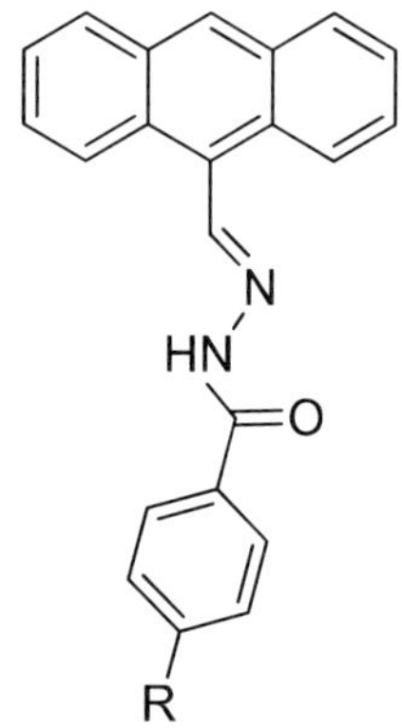

9-Anthraldehyde 4-R-benzoylhydrazone

The thiosemicarbazone derivative of anthracene (**29**) and it's copper(II) complex (**30**) were synthesized and characterized by spectroscopic, electrochemical, and crystallographic techniques. The interaction of **29** and **30** with calf thymus (CT) DNA was explored using absorption and emission spectral methods and viscosity measurements [12]. The investigations revealed a partial-intercalation binding mode. Their protein binding ability was monitored by quenching of tryptophan emission, using bovine serum albumin (BSA) as a model protein. Their cellular uptake, in vitro cytotoxicity testing on the HeLa cell line, and flow cytometric analysis were carried out to ascertain the mode of cell death. The cell cycle analysis indicated that **29** and **30** cause cell cycle arrest in sub-G1 phase.

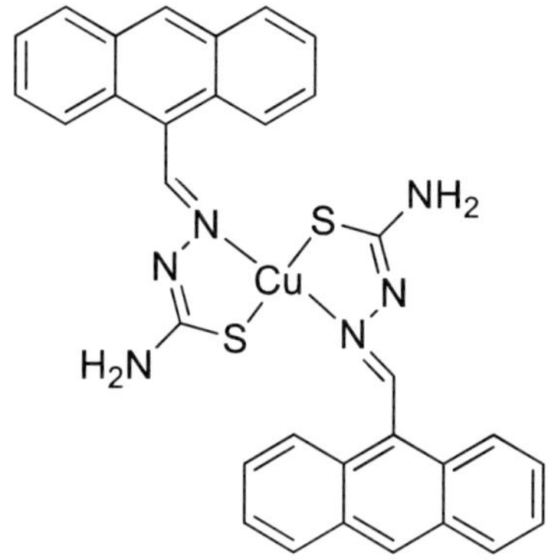

30

The DNA binding and cytotoxicity of Schiff bases **31** and **32** were explored to assess their efficiency in fighting cancer and overcoming the multidrug resistance (MDR) [13]. The **31** and **32** are claimed to treat extremely chemoresistant MCF-7/ADR cell as a drug-sensitive cell, with drug resistance index (DRI), 2.13, showing high potential in overcoming the MDR. The apoptotic ratio induced by **31** and **32** was low for both MCF-7 and MCF-7/ADR cells. The **31** and **32** induced impairment of cell cycle progression of MCF-7 and MCF-7/ADR cell lines and suppressed cell growth by perturbing progress through the G0/G1 phase, the **32** causing more profound effect, which might account for lower drug resistance after treatment with **32**. The molecular docking revealed weak interaction between **31/32** and P-glycoprotein (P-gp), the most important drug efflux pump and intracellular Rhodamine 123 accumulation indicated that the activity of P-gp was not inhibited by **31** and **32**. The combined cellular uptake results have implied that **31** and **32** could bypass P-gp efflux to exert anticancer activity. Anthracene-based Schiff base **33** was synthesized via a one-pot reaction and utilized as a fluorescence turn-on sensor for Fe^{3+} ions, in THF via chelation-enhanced fluorescence at pH 2–14 [14]. The 2:1 stoichiometry of the sensor complex was inferred from Job plots based on UV-Vis absorption titrations. In addition, the binding sites of sensor complex were established from the ^{1}H NMR titrations and supported by the fluorescence reversibility by adding metal ion and pentamethyl diethylene triamine sequentially. The detection limit (LOD) was calculated as 2.95×10^{-6} M and association constant (Ka) 1.88×10^{5} M^{-1}.

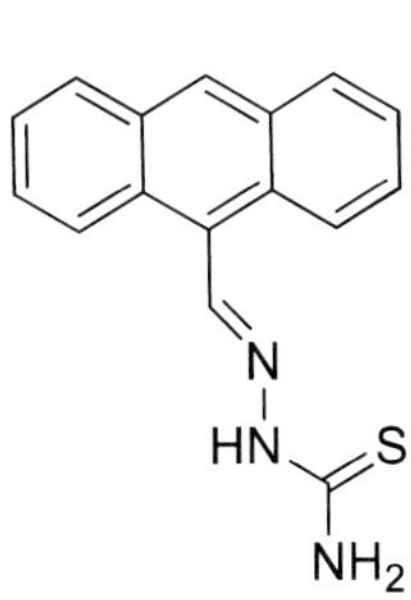

31

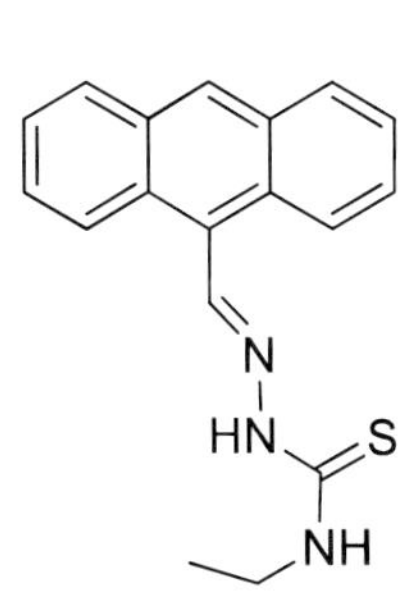

32

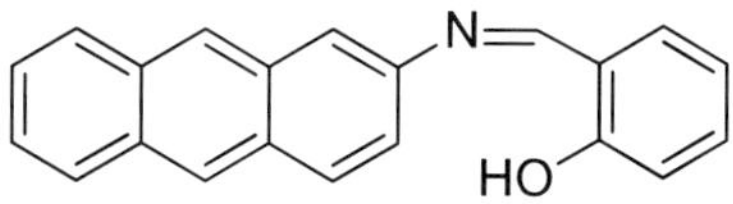

33

Three new anthryl Schiff bases (**34-36**) of similar molecular structure [15] were synthesized through a simple method (Scheme 8). Among them, the **34** displayed an aggregation-induced emission (AIE) feature, **35** exhibited an aggregation-induced emission enhancement (AIEE) property, while **36** showed aggregation-caused quenching (ACQ) behavior. The large conjugation and the existence of multiple intra- and intermolecular interactions restrict the intramolecular vibration and rotation, which benefit the emission in the condensed state, while the tight dimeric structure and intramolecular torsional motion induce fluorescence quenching. The **35** is utilized as fluorescence on–off type sensor for Cu^{2+} in methanol–H_2O (4/1, v/v, pH 7.2: HEPES buffer) and Zn^{2+} in pure methanol. The 2:1 ligand-to-metal coordination patterns of both the metal ions were deduced from a Job's plot, and further confirmed by X-ray crystal structures of the complexes. The **35** has shown a piezofluorochromic characteristic.

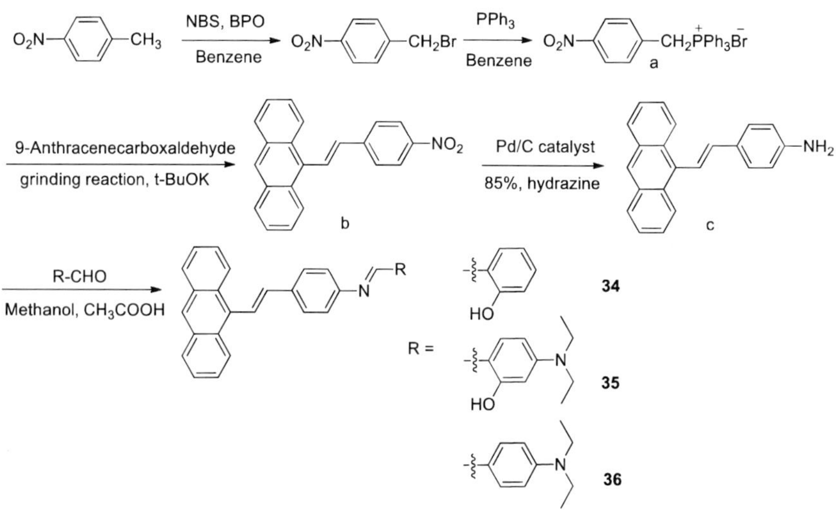

Scheme 8.

A series of complexes of the type $[Cu(L)(phen)(CH_3CN)](ClO_4)_2$ where L = N^1-(anthracen-10-ylmethylene)-N^2-methylethane-1,2-diamine (**37**), N^1-(anthracen-10-ylmethylene)-N^2,N^2-dimethylethane-1,2-diamine (**38**), N^1-(anthracen-10-yl-methylene)-N^2-ethylethane-1,2-diamine (**39**), N^1-(anthracen-10-ylmethylene)-N^2,N^2-diethylethane-1,2-diamine (**40**) or N^1-(anthracen-10-ylmethylene)-N^3-methylpropane-1,3-diamine (**41**) and phen = 1,10-phenanthroline, were synthesized and characterized by spectral and analytical methods [16].

The X-ray crystal structure of the complex of **41** has revealed that the coordination geometry around Cu(II) is square pyramidal distorted trigonal bipyramidal (τ, 0.76). The corners of the trigonal plane of the geometry are occupied by nitrogen atoms. The K_b values calculated from absorption spectral titration of the complexes ($\pi \rightarrow \pi^*$, 250–265 nm) with Calf Thymus (CT) DNA vary in the order **41** > **40** > **39** > **38**> **37**. The fluorescence intensity of the complexes (520–525 nm) decreases upon incremental addition of CT DNA, which reveals the involvement of phen rather than the appended anthracene ring in partial intercalation with the DNA base stack. The complex of **41** interacts with DNA more strongly than the species **40** due to stronger involvement in hydrophobic DNA interaction of the anthracenyl group, which is facilitated by the propylene ligand backbone with chair conformation.

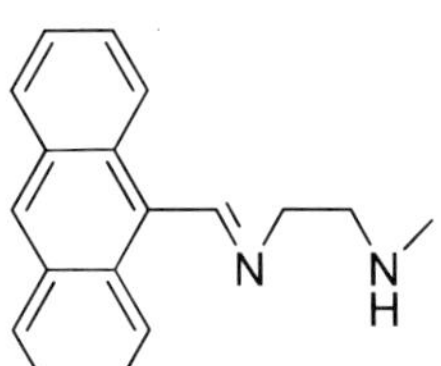

37

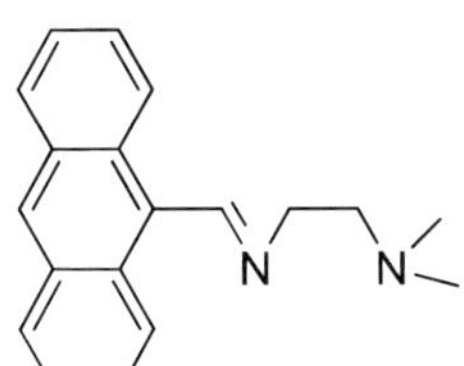

38

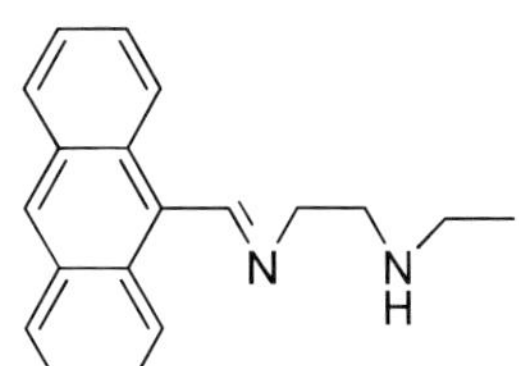

39

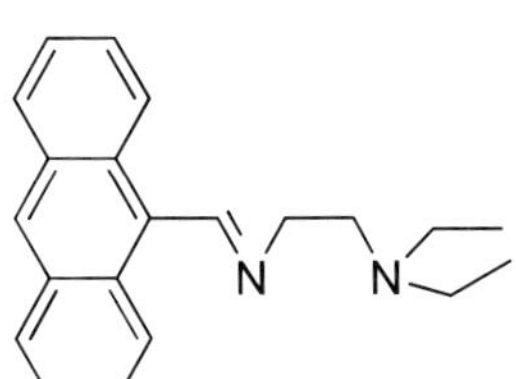

40

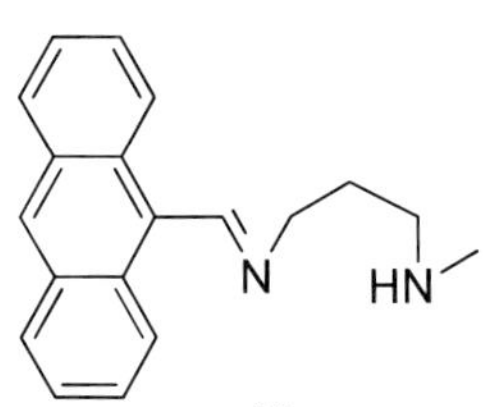

41

The ability of complexes (100 μM) to cleave DNA (pUC19 DNA) in 5 mM Tris-HCl/50 mM NaCl buffer at pH 7.1 in the absence of a reducing agent or light varies in the order **41 > 40 > 39 > 38> 37**, which is in conformity with their DNA binding affinities. Interestingly, cytotoxicity studies on the MCF-7 human breast cancer cell line show that the IC50 value of the complex of **41** is less than that of cisplatin for the same cell line, revealing that it can act as an effective cytotoxic drug in a time-dependent manner.

Three anthracene-based Schiff bases **42-44** (**42** = (E)- *N′*-((anthracen-10 - yl)methylene) benzohydrazide; **43** = (E)-1-((anthracen-10-yl) methylene)-4-phenylsemicarbazide; and **44** = (E)-1-((anthracen-10-yl) methylene)-4-phenyl-thiosemicarbazide) were synthesized from 9-anthracenecarboxaldehyde, benzohydrazide, 4-phenylsemicarbazide and 4-phenylthiosemicarbazide respectively, and characterized by various spectral techniques [17]. The absorption spectral characteristics of **42-44** were bathochromically tuned to the visible region by extending the π conjugation. These target compounds were found weakly fluorescent in solution made in tetrahydrofuran (THF) because of rapid isomerization of the C=N double bond in the excited state.

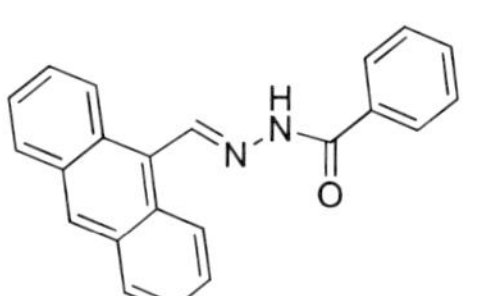

42

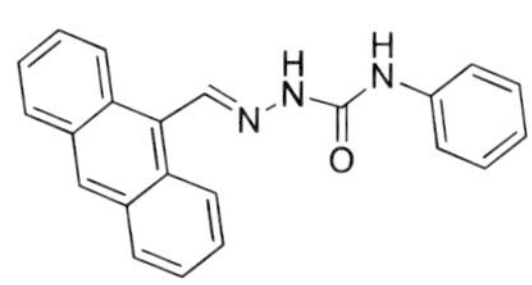

43

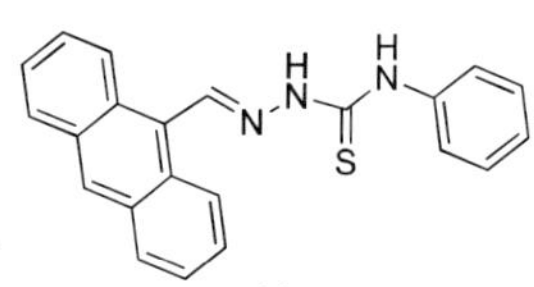

44

However, gradual increase of water up to 90% in the dispersion of **42-44** made in THF/water mixture has resulted in an increase in the fluorescence intensity mainly due to aggregation - induced emission enhancement (AIEE) properties. The formation of nano-aggregates of **42-44** was confirmed by scanning electron microscopy (SEM) and atomic force microscopy (AFM). The compounds **42-44** were found good probes for the fluorescence sensing of bovine serum albumin (BSA) and breast cancer cells by optical cell imaging. Anthracene based Schiff bases **45** and H(**46**) and their Cu(II) complexes [Cu(**45**)Cl_2] and [Cu(**46**)Cl] (where **45** = *N*[1],*N*[2]-bis(anthracene-9-methylene)benzene-1,2-diamine, H(**46**) = (2Z,4E)-4-(2-(anthracen-9-ylmethylene amino)phenylimino)pent-2-en-2-ol) were synthesized and characterized by elemental analysis, spectral techniques and cyclic voltammetry [18]. The electronic structures and geometrical parameters of the complexes were analyzed by theoretical B3LYP/DFT method. The interaction of these complexes with CT-DNA was explored by absorption, cyclic voltammetric and CD spectral studies. From the electronic absorption spectral studies, the DNA binding constants of complexes were found as 8.7×10^3 - 7.0×10^4 M^{-1}. From electrochemical studies, the ratio of DNA binding constants K_+/K_{2+} for the complex of **46** has been estimated as > 1. The high binding constant values, K_+/K_{2+} ratios greater than unity and a positive shift in $E_{1/2}$ value in the voltammetric titration of DNA with the complex of **46** suggest that it binds more avidly with DNA than the complex of **45**. The inability to affect the conformational changes of DNA in the CD spectrum has supported electrostatic binding by the complex of **45**. Probably bulky anthracene unit sterically inhibited these complexes from intercalation and thereby their binding was in the groove or electrostatic. The complex of **46** has hardly cleaved supercoiled pUC18 plasmid DNA in the presence of H_2O_2. The complex of **46** probably binds to DNA through minor groove binding.

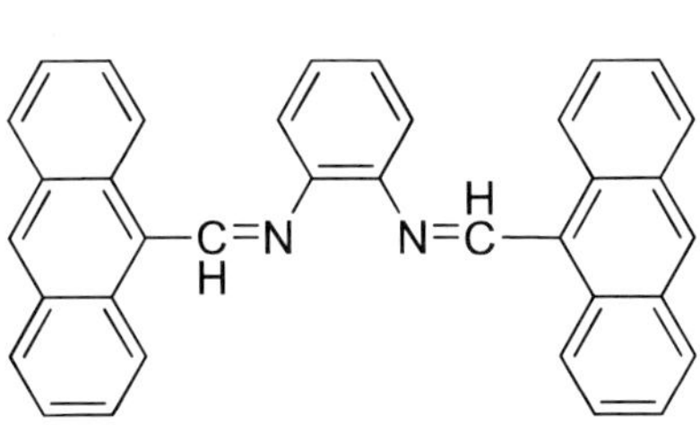

45

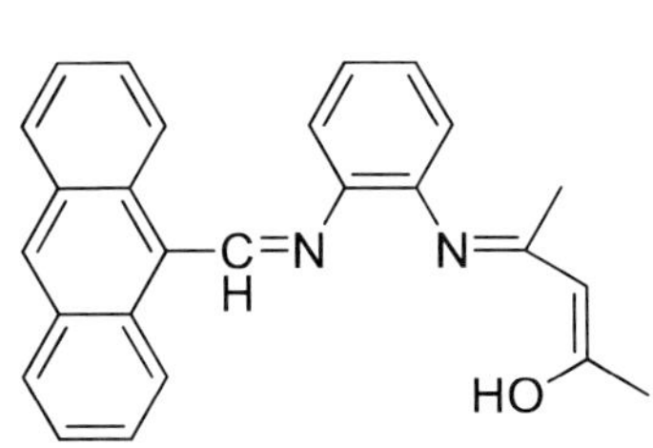

46

The complexes [Cu(**47**)Cl_2], [Ni(**47**)Cl_2], [Cu(**48**)Cl_2] and [Ni(**48**)Cl_2]), where **47** = (1-(anthracen-9-yl)-*N*-(1H-benzo[d]imidazole-2-yl) methanimine) and **48** = (1-(anthracen-9-yl)-N-(benzo[d]thiazol-2yl) methanimine) were synthesized and characterized by various spectral and electroanalytical techniques [19]. The molecular structures of complexes of **47** and **48** were determined by the single crystal X-ray diffraction. The absorption and emission spectral approaches have revealed that these complexes may be partially stacking between the DNA bases. In contrast, the possibility of their covalent interaction was evaluated with the cyclic voltammetric techniques. All the results revealed that the interaction of four complexes with HS-DNA was via covalent/partial intercalation mode and the quenching mechanism of BSA by all these complexes is through static mode. In addition, in vitro cytotoxicity of the complexes on human gastric cancer cell lines were observed by MTT and the fluorescence based morphological assessment obtained using Hoechst 33342 staining. It revealed that these complexes induced apoptosis against AGS cancer cells. The best result was obtained with a Ni complex of **48** (8.9 mM).

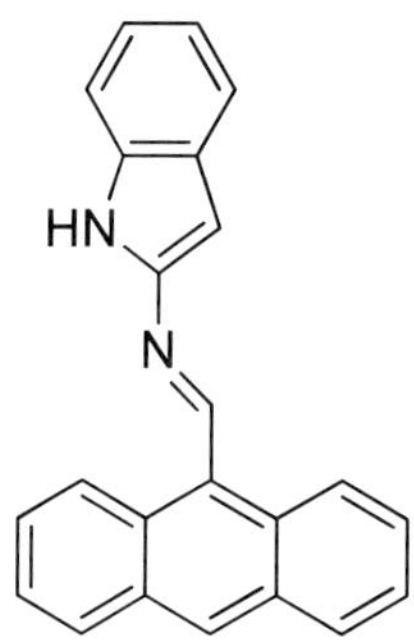

47

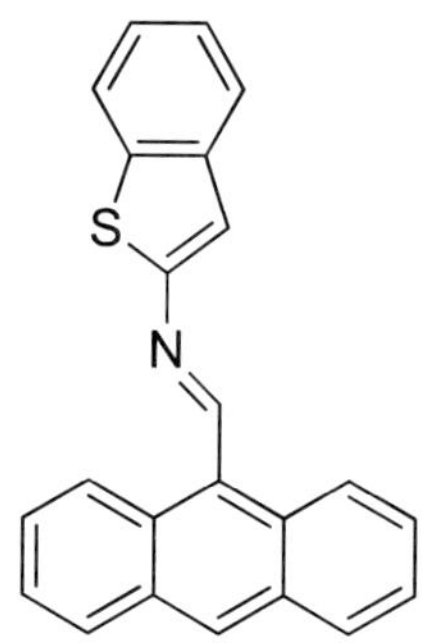

48

Three non-conjugated methylene–anthracene Schiff base molecules (**49-51**) were synthesized [20]. The **49** and **51** showed mechano-fluorochromic properties, and distinctive 26 and 37 nm hypochromic shifts respectively. The experiment results revealed that there was no specific relationship between alkoxy chain lengths and their stimuli–response behavior. The PXRD profiles demonstrated a transformation from crystalline to the amorphous state upon grinding.

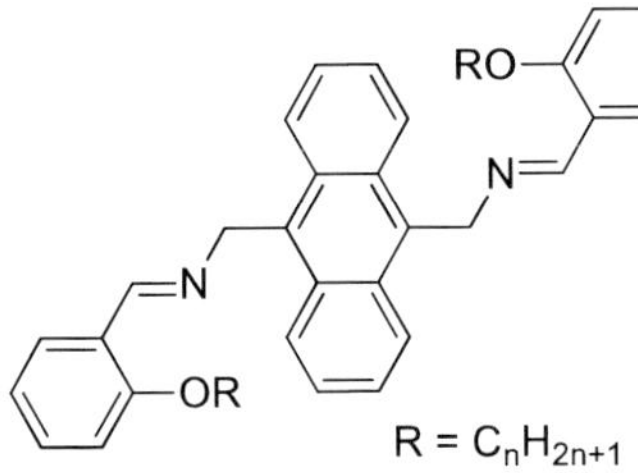

R = C_nH_{2n+1} n = 4; **49**
n = 8; **50**
n = 12; **51**

The crystal and molecular structures of Ni(II) complexes with two tridentate anthracene containing Schiff bases prepared by condensation of 9-anthraldehyde with diethylenetriamine **(52)** and with dipropylenetriamine **(53)** were reported [21]. The complexes were synthesized by reacting $Ni(OAc)_2{\cdot}4H_2O$ with the ligands in 1:1 molar ratio.

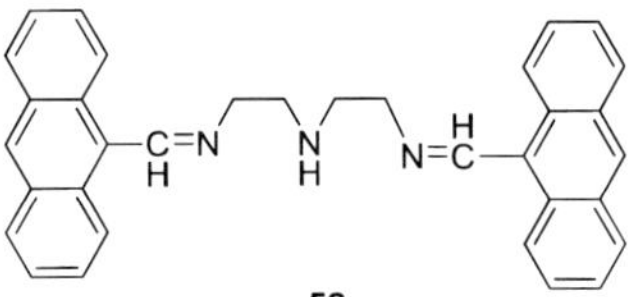

52

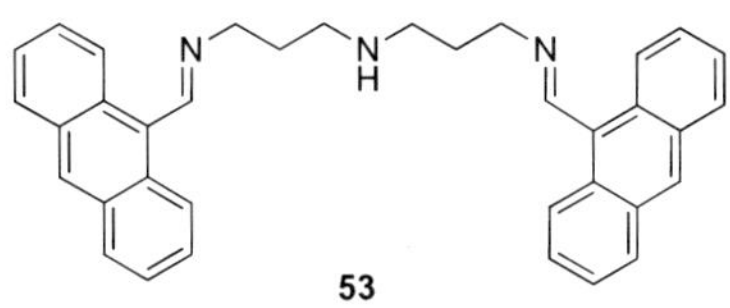

53

N^2, N^3-Bis(anthracen-9-ylmethylene)pyridine-2, 3-diamine (**54**), synthesized via condensation of 2,6-diaminopyridine with anthracene-9-carbaldehyde was used to prepare its praseodymium (Pr), erbium (Er), and ytterbium (Yb) complexes. FT-IR spectral studies revealed bidentate nature of **54** in these complexes. The cytotoxicity activity versus Vero, human breast cancer (MCF7), and cervical (HeLa) anticancer cell lines were studied.

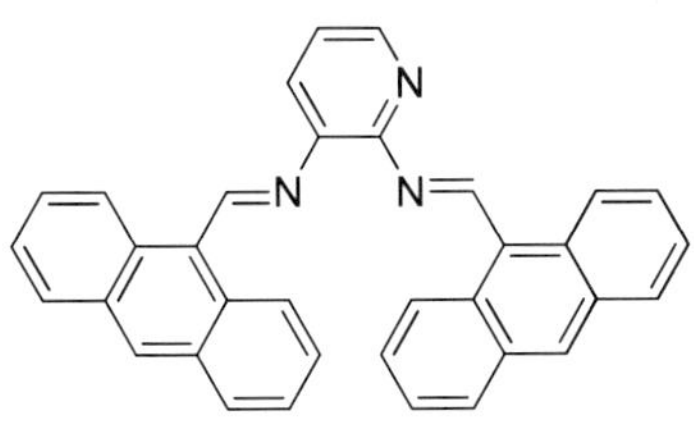

54

5-[1-(Anthracen-9(10H)-ylideneamino)-2-(1H-imidazol-5-yl)ethyl]-1,3,4-thiadiazol-2-amine(**55**) synthesized via condensation reaction of *N*-anthracen-9(10H)-ylidene histidine with thiosemicarbazide in $POCl_3$ was characterized by FT-IR spectroscopy, electronic spectroscopy, ^{1}H NMR and $^{13}C\{^1H\}$ NMR. In *vitro* tests for fungicidal activity show that **55** displays good activity against *Gibberela*, *Cercospora arachidicola*, *Physolospora piricola* and *Fusarium oxysporum.*

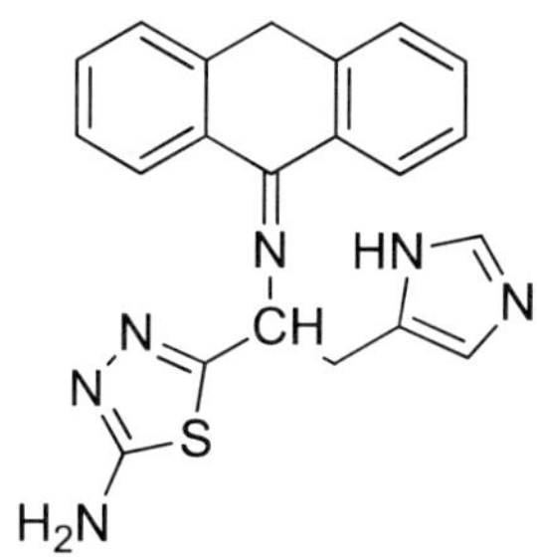

55

Schiff-bases derivatives have been synthesized from 3,5-diaryl-4-amino-1,2,4-triazole (**56**) and 9-anthraldehyde. The emission spectrum of 3,5-diphenyl-4-(anthrylmethyleneamino)-1,2,4-triazole exhibited fluorescent quenching upon chelation to Fe^{3+} ions [24a].

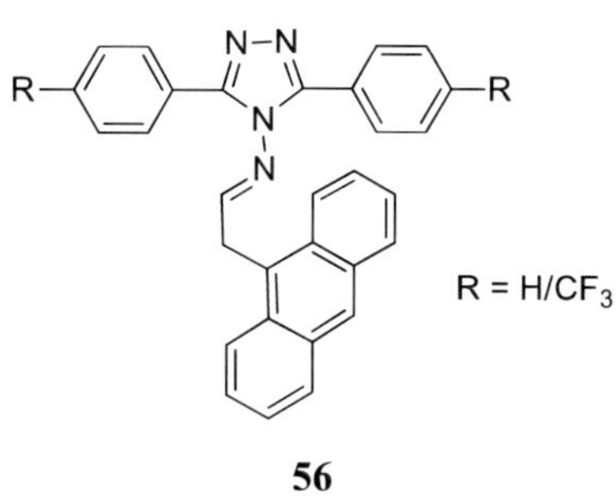

56

Anthracene and pyridine-containing Schiff base derivative 2-(2-(anthracen-9-ylmethylene)-hydrazinyl) pyridine (**56'**) was synthesized via one-pot reaction[24b]. It worked for fluorescent ''off–on'' detection of Cu^{2+}

ions, via PET-based mechanism. The 1 : 1 stoichiometry of **56'**-Cu^{2+} sensor complex calculated from Job's plot based on PL titrations was supported by ESI (+ve) mass spectral analysis. The binding sites in **56'**···Cu^{2+} sensor conjugate were supported by ^{1}H NMR titration. The detection limit (LOD) and association constant (K_a) of **56'**-Cu^{2+} complex were obtained by standard deviation and linear fittings. The quantum yield (ϕ_f), SEM analysis, pH effect, FT-IR and density functional theory (DFT) studies are reported for the sensing conjugate. More decisively, confocal fluorescence microscopy imaging from Raw 264.7 cells indicated that **56'** could be used as an effective fluorescent probe for the analysis of Cu^{2+} ions in living cells.

AMINES

Diaminoanthracene derivatives, 9,10-bis(1-naphthylphenylamino) anthracene (**57**), 9,10-bis(2-naphthylphenylamino)anthracene (**58**), 9,10-bis(*m*-tolylphenylamino)anthracene(**59**), and 9,10-bis(diphenylamino) anthracene (**60**) were conveniently synthesized from the corresponding diarylamine and 9,10-dibromoanthracene in the presence of Pd-$(OAc)_2$, tri-*tert*-butylphosphine, and sodium *tert*-butoxide in *o*-xylene [25]. Electroluminescent devices using **57, 58**, and **60** as the hole transporters and host emitters were designed. The devices were shown to emit intensive green light from the diaminoanthracene layer instead of the Alq_3 layer. Very high brightness, current, and power efficiency and excellent CIE coordinate were observed.

9,10-Bis(di(p-*t*-butylphenyl)amino)anthracene (**61**) was employed [26a] as a photocatalyst for radical fluoroalkylation under visible light irradiation (Scheme 9). The diarylamine (donor)−anthracene (π conjugated system)−diarylamine(Donor) scaffolds were easily accessed by typical palladium-catalyzed cross-coupling protocols of the corresponding halogenated anthracenes with various lithium diarylamides (Scheme 9). The anthracene-based photocatalyst exhibited high reducing power, leading to generation of versatile fluoroalkyl radicals such as tri- and difluoroethyl/methyl from the corresponding electron-accepting precursors.

The catalyst design strongly influenced the absorption capability of visible light and the stability toward redox stimuli.

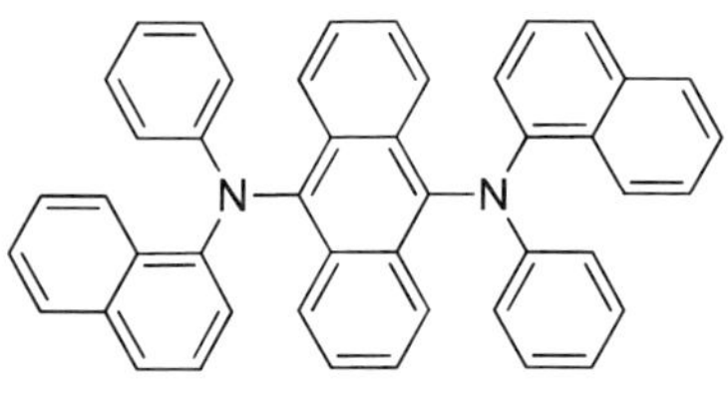

57

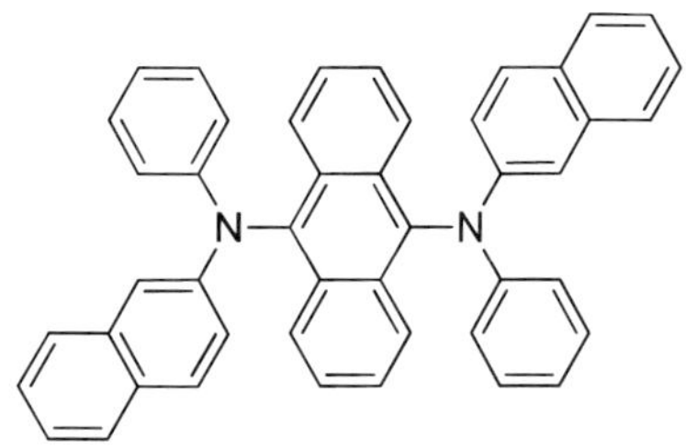

58

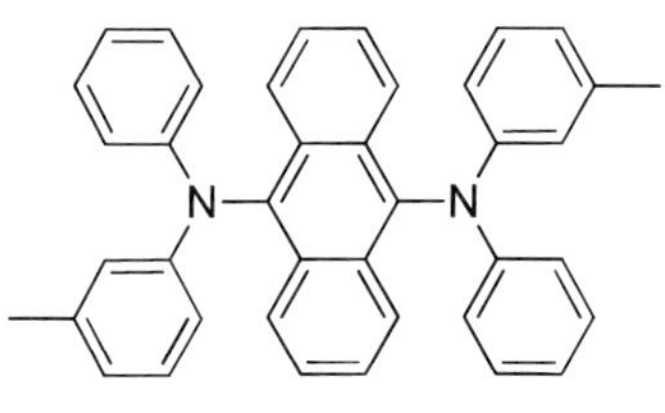

59

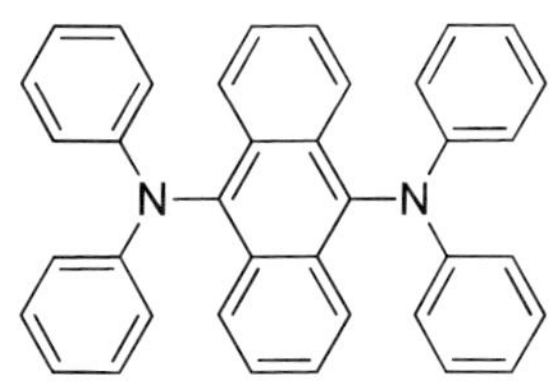

60

The detailed mechanistic studies on the metal-free photocatalytic amino-trifluoroethylation of styrene with diphenyl(2,2,2-trifluoroethyl) sulfonium trifluoromethanesulfonate suggested that the reaction proceeds via catalytic radical process instead of radical chain process. In addition, the static quenching process was involved in the first single-electron-transfer from the photocatalyst to the fluoroalkylating reagent. Moreover, the 1e-oxidized cationic radical species of **61**, a key active catalytic species with long lifetime and a characteristic 'Intervalence Charge Transfer Band' in the near IR region, was detected. From the viewpoint of elemental strategy initiative and green chemistry, this noble metal-free organic photocatalytic system is claimed to have the potential to provide a pivotal technology to replace ruthenium- and iridium-based metal photocatalysis.

The Cu(II) complexes of anthracene-based amine ligands, **61a** and **61b**, were reported. The **61b** was found to be relatively robust and competent in catalytic O_2 reduction than **61a** [26b].

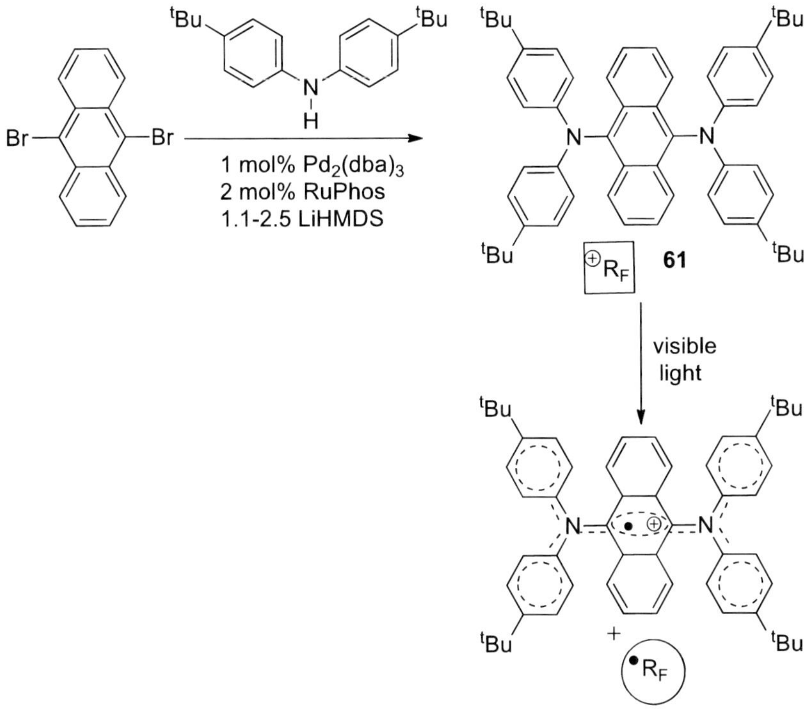

Scheme 9.

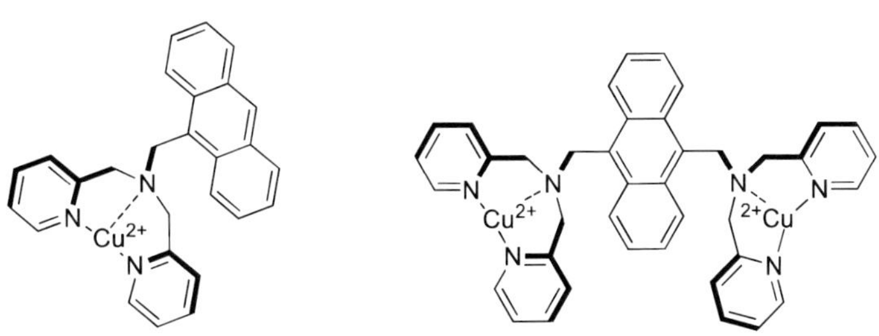

61a **61b**

N-Heterocycle Bearing Ligands

The fluorescent 9,10-bis(4-pyridyl)anthracene (**62**) was reacted with β-diketonate metal complexes of Mn(II), Zn(II), Co(II) and Cu(II) [27]. The resulting dinuclear and polynuclear complexes were found to show luminescent properties diminished relative to that of **62,** except in the case of Zn(II), which showed an enhancement.

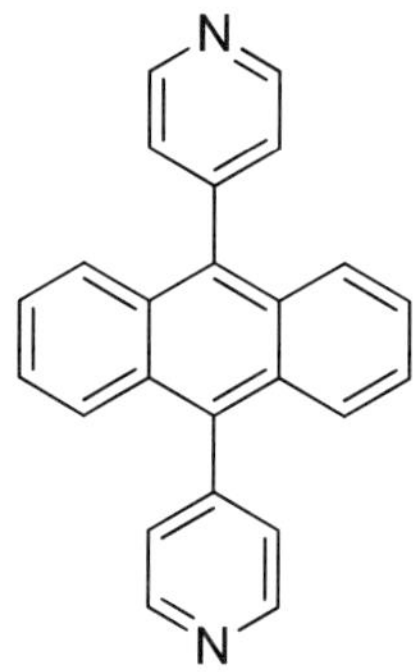

62

The self-assembly of (TMEDA)Pd$(NO_3)_2$ or (TMEDA)Pt$(NO_3)_2$ (where TMEDA = N^1,N^1,N^2,N^2-tetramethylethane-1,2-diamine) and anthracene based diimidazole ligands (**63-64**) in aqueous solution afforded [28] a series of positively charged $[M_2L_2]^{4+}$ (L = **63-64)** dimetallomacrocycles. Their structures were authenticated by ^{1}H NMR and electrospray ionization mass spectrometry and in the cases of {[(TMEDA)Pd]$_2$(**63**)$_2$}$(NO_3)_4$, and {[(TMEDA)Pd]$_2$(**63**)$_2$}$(PF_6)_4$ by single-crystal X-ray diffraction analysis. The interactions between metallomacrocycles and various anions were investigated via fluorescence titration and cyclic voltammetric studies. The fluorescence titration experiment exhibited the formation of 1:1 host-guest complex for anthracene-based positively charged $[M_2L_2]^{4+}$ (L = **63-64)** metallomacrocycles with NO_3^-.

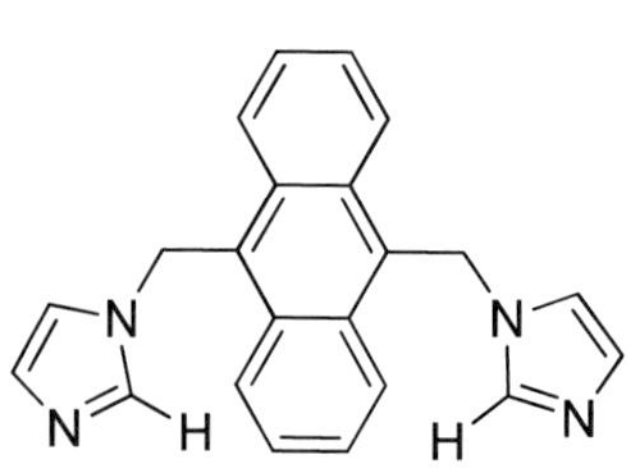

63

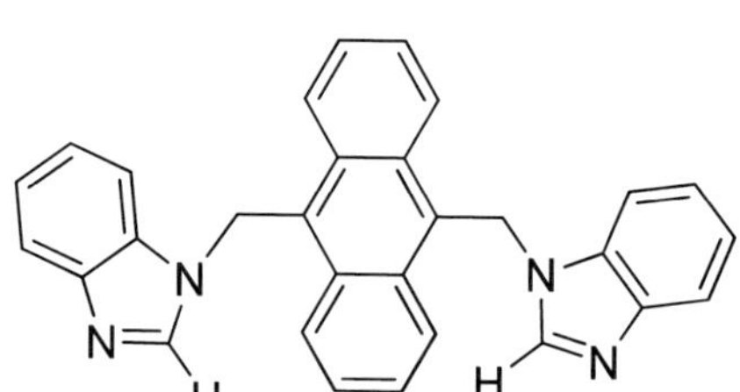

64

The combination of linear dipyridyl ligands with a modular building unit, based upon a 1,8-platinum-functionalized anthracene, led to the self-organization of rectangular framework (**65**). The X-ray crystallography has confirmed the cyclic structure of the supramolecular cationic complex [29].

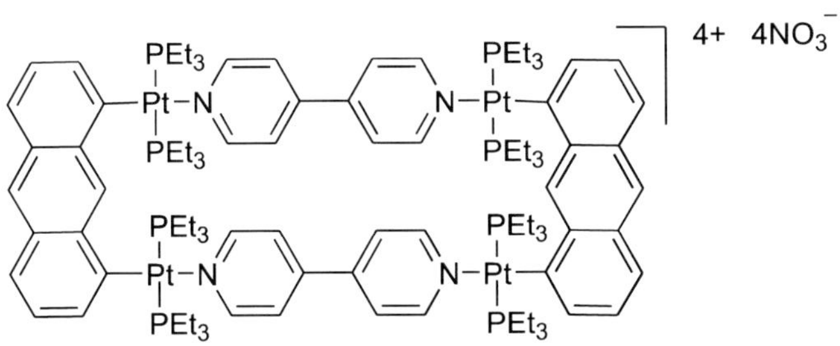

65

The adsorption of the dye 4-(7-{10-[bis(4-hydroxy-phenyl)-amino]-anthracen-9-ylethynyl}-2-octyl-2H-benzotriazol-4-yl)-benzoic acid (**66**) and with a co-adsorbent chenodeoxycholic acid (CDCA) on the anatase $TiO_2(001)$ surface was investigated using density functional theory [30] as a crucial component of a high-performance dye-sensitized solar cell. The results showed an active role CDCA co-adsorbent not only in stabilizing the dye/oxide system but also in preventing the surface tensile stress induced by the dye monolayer. The monolayer of the dye **66** at high surface coverage in the absence of the co-adsorbent applies tensile stress to the (001) surface of TiO_2 due to interactions of repulsive type between the polar **66** molecules. The induced tensile stress shift both the valence band and the conduction band to higher energy, which reduces the energy gap between the lowest unoccupied molecular orbital (LUMO) of **66** and the conduction band maximum of $TiO_2(001)$. The reduced LUMO−conduction band maximum gap is detrimental to the charge injection efficiency of the dye-sensitized solar cell, which results in a loss of power conversion efficiency. The co-adsorbent increases the spacing between the dye molecules to reduce the surface tensile stress and thus prevents the decrease in the charge injection efficiency.

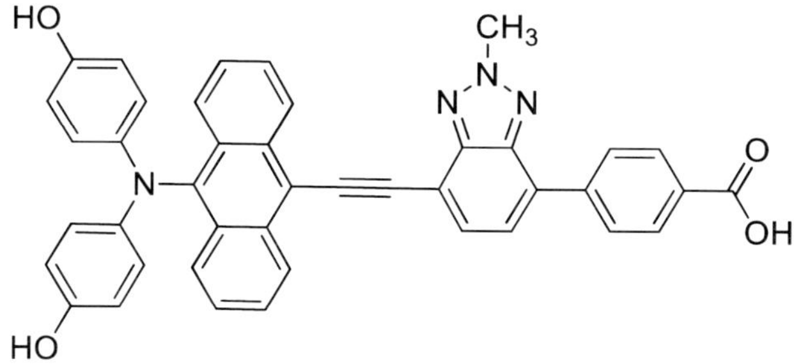

66

The Cu^{2+} ion-induced formation of a water-soluble metallocyclophane (**67**) has been reported. It uniquely recognizes guanosine 50-monophosphate through changes in its emission intensity. These changes are attributed to the synergistic effects of electrostatic, coordinate and π-stacking interactions inside the cavity [31].

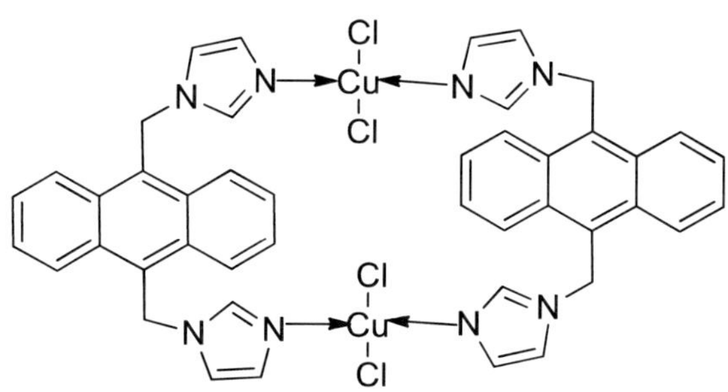

67

Cyclophanes possess a cavity of defined size and are efficient in encapsulating and stabilizing guest molecules inside the cavity through various non-covalent interactions. This unique property of the cyclophanes has been widely exploited for the development of selective probes for a variety of guest molecules. They, including anthracene ones, are reviewed [32] highlighting their use in sensitive and selective optical recognition of important biomolecules, such as DNA and 50-ATP. Anthracene based Schiff-base pyrrole macrocycle (**68**) was synthesized straightforwardly [33]. Its cobalt complex acts as an efficient and selective catalyst in PhCN solution for the four electron reduction of oxygen to water, displaying activity similar to that of the well-known co-facial diporphyrin analogues [34].

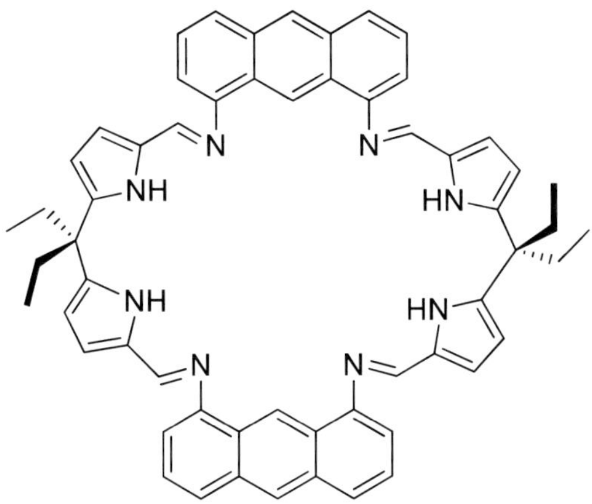

68

Half-sandwich arene complexes of Ru(II) of the type [Ru(η^6-arene)(**L**)Cl](PF_6), where arene is benzene or *p*-cymene, and **L** is 1-(anthracen-10-ylmethyl)-4-methylhomopiperazine(**69**), were isolated and characterized with spectral methods. The single crystal X-ray diffraction studies have revealed that they have a pseudo-octahedral "piano-stool" structure equipped with the arene ligand as the seat and the donor sites of bidentate ligand and the chloride ion as the legs of the stool. Their DNA binding affinities are reported. They cleave DNA effectively even in the absence of an external agent and the cleaving ability is enhanced in the presence of an activator like H_2O_2 [35]. Two new blue materials 4-(10-(9-phenyl-9H-carbazol-3-yl)-anthracen-9-yl)benzonitrile (**70**) and 4-(10-(4-(9H-carbazol-9-yl)phenyl)anthracen-9-yl)benzonitrile (**71**) were synthesized.

These compounds have exhibited a small energy gap and orbit overlapped between the excited state of T_m and S_n, which accelerate the triplet–triplet annihilation process. When these two compounds were used as emitters in non-doped devices, the triplet energy was utilized efficiently through triplet–triplet annihilation; excellent blue electroluminescence performances were achieved. The maximum external quantum efficiency values were 10.06 and 9.23% for **70** and **71**, respectively. The efficiency roll-off for **70** and **71**, were very low and the EQEs remained 8.97 and 7.10% at the luminescence of 1000 cd m^{-2}. (0.14, 0.14) and (0.14, 0.10) [36].

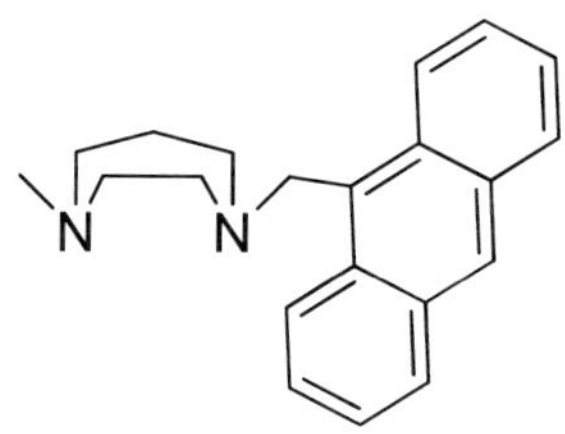

69

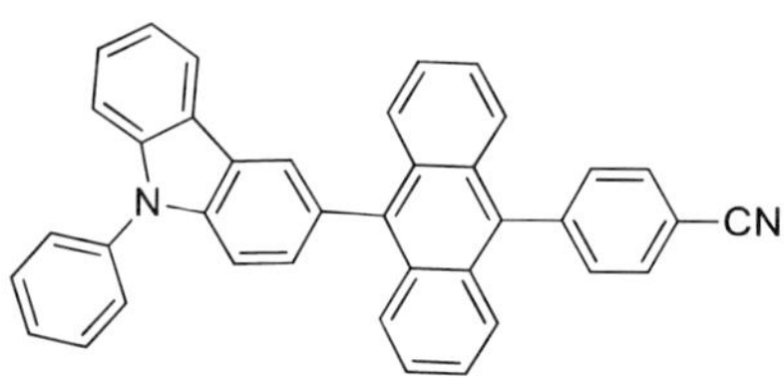

70

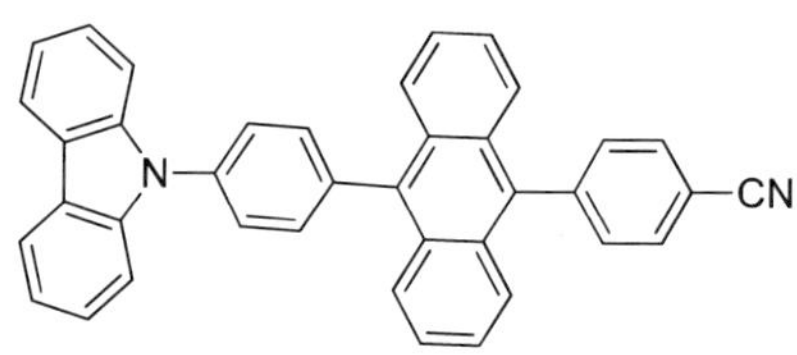

71

The pincer ligand, anthracene-1,8-diylbis(N-3,5-bistrifluormethyl-aniline) (**72**) is reported. On treating $Hf(NMe_2)_4$ with **72** the bimetallic complex[(**72**){$Hf(NMe_2)_3NHMe_2$}$_2$} was formed [37].

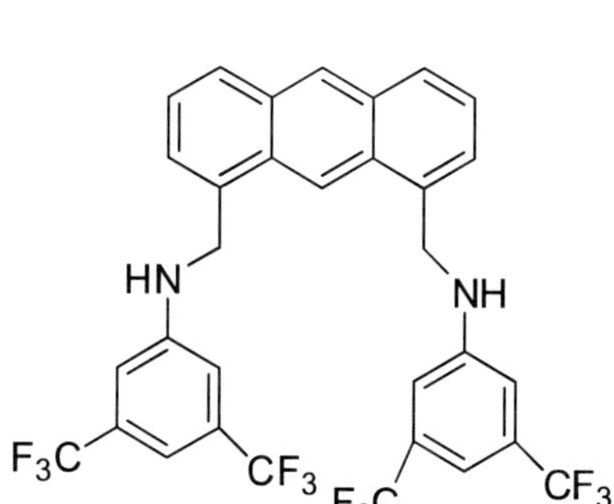

72

Two linear 9,10-anthracene-based small molecules (**73-74**, SMs) with $(D\text{-}A)_2Ar$ framework were synthesized for BHJ-OSCs. Both $(D\text{-}A)_2Ar$-type SMs exhibited better light-harvesting ability and lower HOMOs than the parent D-A-type molecule. The highest PCE of 1.72% was exhibited in the optimized $(DTBTTPA)_2An/PC_{61}BM$ based device with a *J*sc of 6.17 mA cm $^{-2}$, *V*oc of 0.85 V and FF of 33% under simulated AM 1.5G (100mW cm^{-2}) irradiation. These results demonstrate that the constructed $(D\text{-}A)_2Ar$-type SMs sufficiently improve the photo-physical and photovoltaic performances[38].

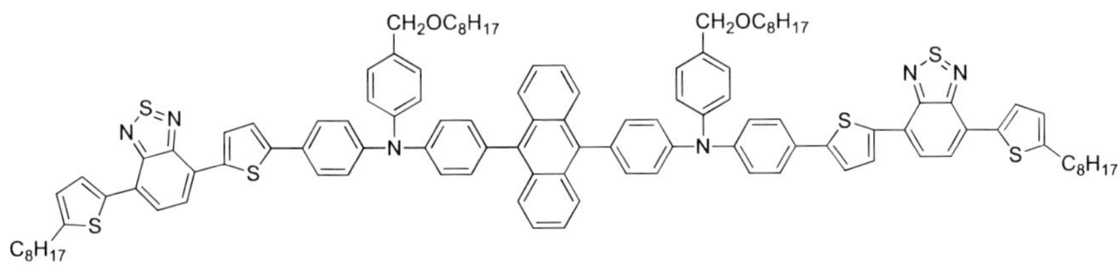

73

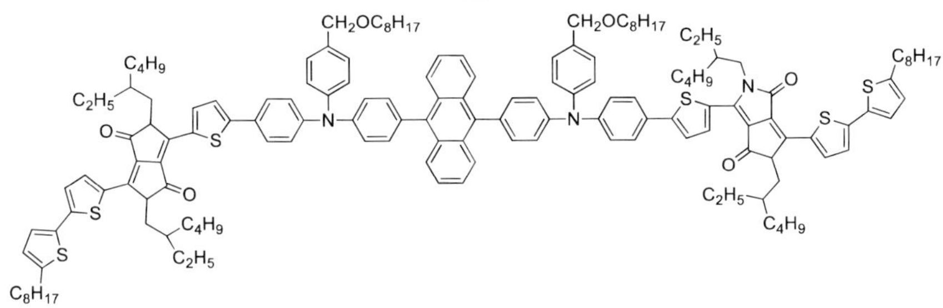

74

Phosphorous Donors

1,8-Bis(dipheny1phosphino)anthracene (**75**) was synthesized from dipotassium 1,8-anthracenedisulfonate and potassium diphenylphosphide. On reaction of **75** with nickel(I1) chloride and bis-(benzonitrile) palladium(II) chloride, cyclometallation of the C-H bond in anthracene at 9-position led to the square-planar chelate complexes

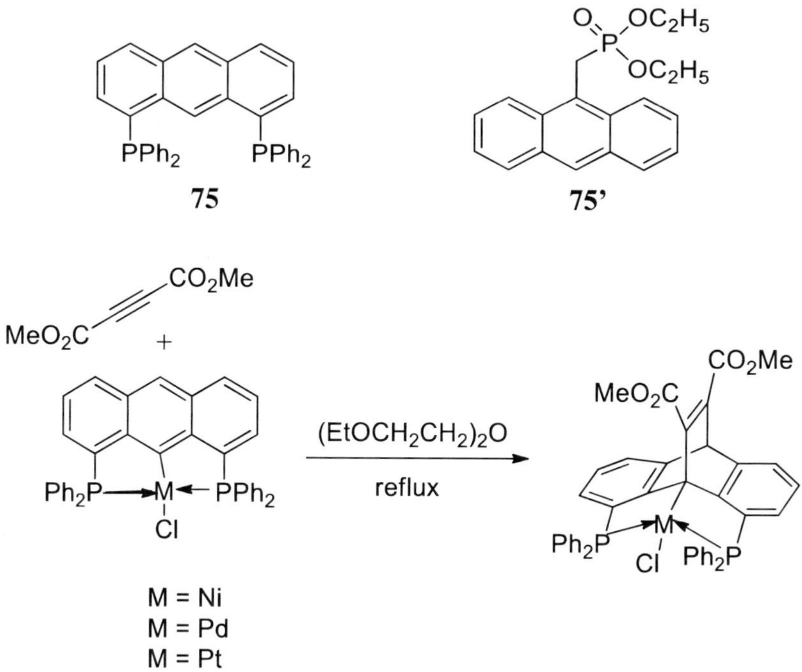

Scheme 10.

Treatment with aqueous potassium cyanide did not remove nickel from its complex but substituted its chloride with cyanide, demonstrating the high stability of the cyclometallated chelates of nickel [39a]. The lanthanide complexes, [Ln(hfac)$_3$(**75'**)(H_2O)] [Ln(III) = Dy / Gd], [Dy(hfac)$_3$(**75'**)$_2$]$_2$·H_2O and [Gd(hfac)$_3$(**75'**)$_2$]·2H_2O, were synthesized (hfac = hexafluoroacetylacetonate), by using one or two **75'** molecules to substitute coordination water molecules of Ln(hfac)$_3$(H_2O)$_2$. The number of introduced **75'** ligands modified the coordination geometry of Ln(III) ions, showing a distorted biscapped triangular prismatic geometry in iso-structural first two complexes and a distorted square antiprismatic geometry in other two. The magnetic studies have revealed that both first and third complexes show field-induced slow magnetic relaxation under the applied dc field of 1000 Oe. The solid-state fluorescence measurements have indicated the presence of multicomponent emissions in Dy complexes, including ligand-centered (LC) emissions from hfac and yellow emission from Dy(III) ions and only ligand-centered (LC) emissions in Gd complexes. The metal complexes of **75** undergo [4+2] addition as shown in Scheme 10 resulting dibenzobarrelene-based C(sp^3)-metalated pincer complexes [40].

The thermal reaction of $[Ru_2(CO)_6(l\text{-}PFu_2)(\mu\text{-}\eta^1,\eta^2\text{-}Fu)]$ (Fu = 2-furyl) with (9-anthracenyl) diphenylphosphine (**76**) has been reported to give $[Ru_2(CO)_5(l\text{-}PFu_2)(\mu\text{ - }\eta^1,\ \eta^1,\ \eta^2\text{-}C_{14}H_8PPh_2)]$ in good yield whereas the corresponding reaction between $[(\mu\text{-}H)_4Ru_4(CO)_{12}]$ and **76** gives $[HRu(CO)_3(PPh_2C_{14}H_8)][(\mu\text{-}H)_4Ru_4(CO)_{11}(\mathbf{76})]$. X-ray crystal structures of both compounds are reported [41].

CARBENES

Anthracene derivatives of carbene precursors have been found attractive as receptor for a variety of species. Anthracene bearing imidazolium moieties (**77-79**) were synthesized and reported to show selective binding for pyrophosphate and phosphate [42]. These receptors for the two anions have potential of sensing them by the change in the fluorophore intensity.

A receptor based on anthracene and imidazole building blocks was synthesized (Scheme 11) for selective recognition of terephthalate. The receptor enclosed and acted on the special guest by synergistic effects of cavity size, π–π stacking and hydrogen bonding interaction [43a].

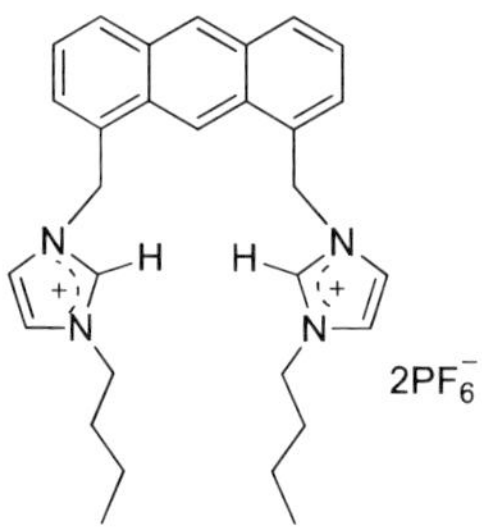

77

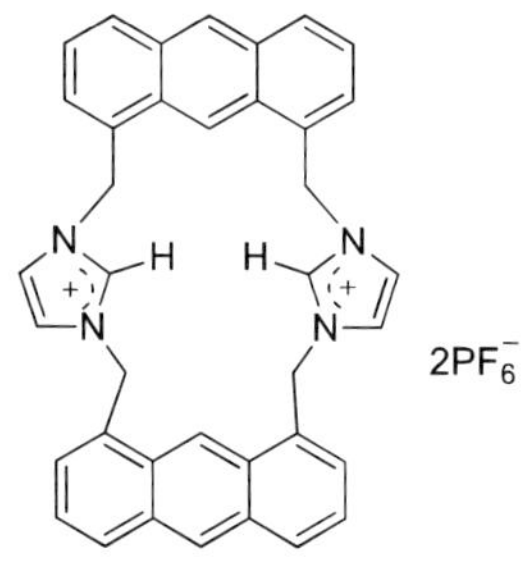

78

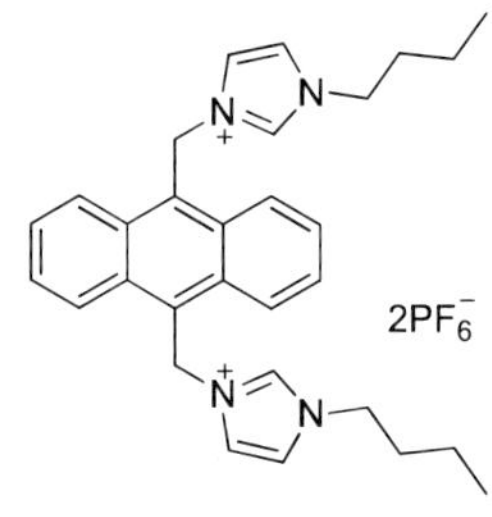

79

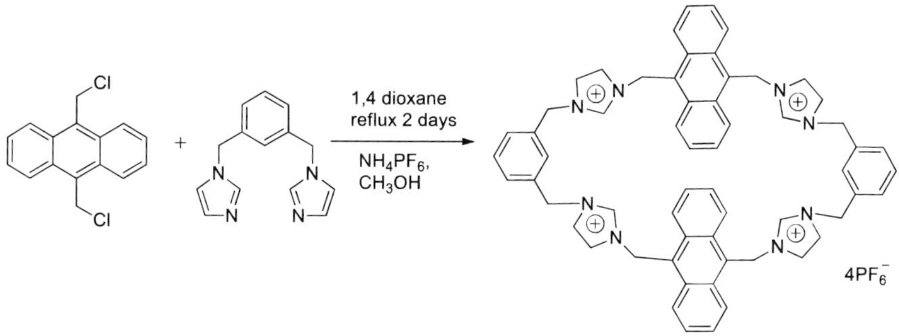

Scheme 11.

The complexes **79b** were synthesized and found efficient as dihydroalkoxylation catalysts [43b]. They show the bimetallic cooperativity.

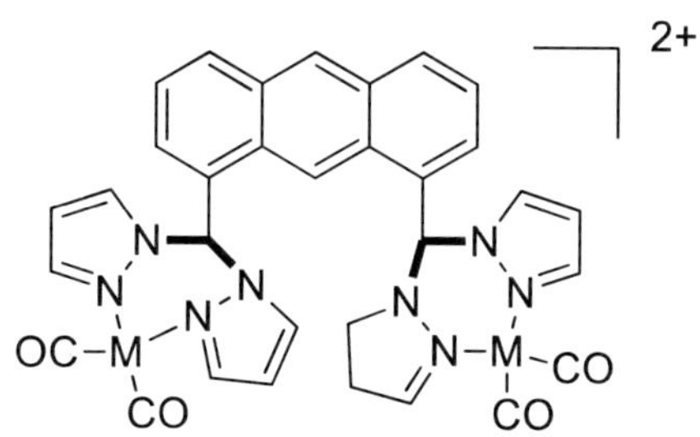

M = Rh/Ir

79b

The *o*-phenylenediamine-based fluorescent cleft **80** was designed [44]. The **80** selectively recognizes tetrabutylammonium dihydrogen phosphate in CH_3CN by exhibiting a significant decrease in the emission of anthracene. In aqueous CH_3OH, no measurable change in the emission is observed for tetrabutylammonium dihydrogen phosphate. On contrary sodium salts of phosphate, hydrogen phosphate, and dihydrogen phosphate exhibit moderate changes. Tetrabutylammonium hydrogen sulfate is sensed effectively in aqueous CH_3OH.

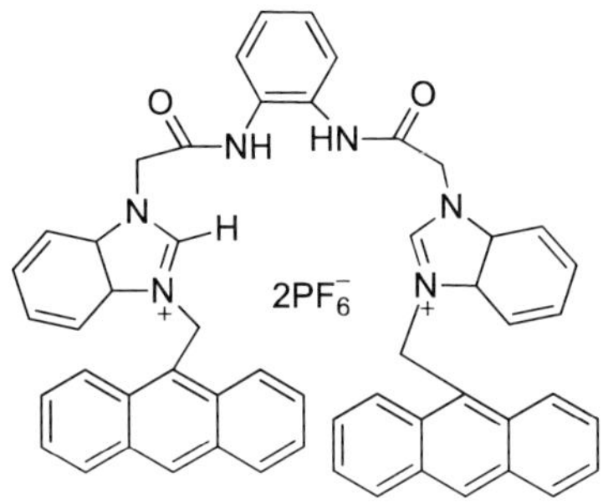

80

A fluorescent chemosensor bearing two imidazolium groups and two anthracene groups (**81**) has been reported [45] for $H_2PO_4^-$. It has displayed a highly selective fluorescence quenching effect and an unique excimer peak only with this anion. The association constant of **81** with $H_2PO_4^-$ was calculated as > 10^6 M^{-1}.

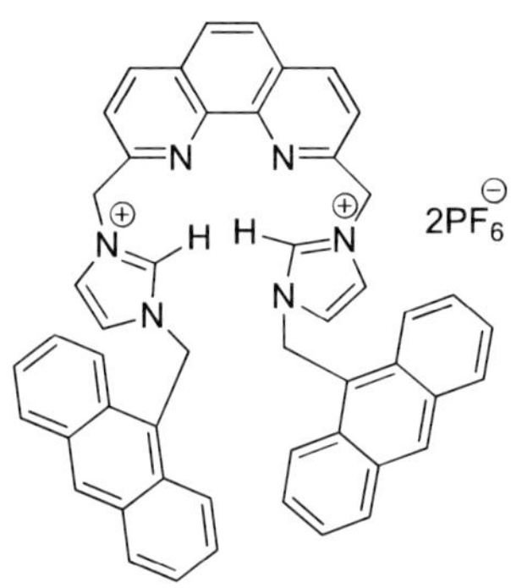

81

Two *o*-phenylenediamine based new receptors **82** and **83** with an anthracene-coupled benzimidazolium motif were synthesized [46]. The directed hydrogen bonds and charge–charge interactions allowed the open clefts of both **82** and **83** to bind carboxylate, fluoride and dihydrogenphosphate anions forming 1:1 complex with moderate binding constant values. The binding cleft of **83** is found to be more effective than that of **82.**

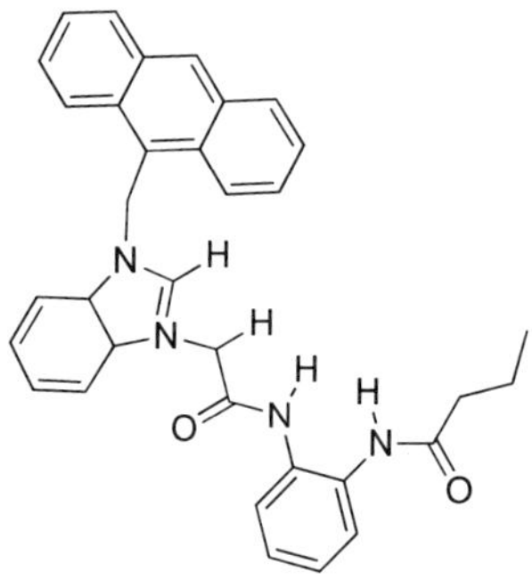

82

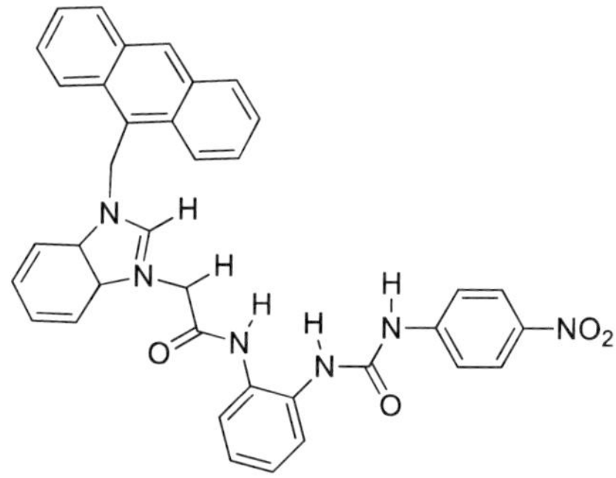

83

1-(9-Anthracenylmethyl)-3-ethylimidazolium iodide (**84**) was treated with $PdCl_2$ and $Pt(cod)Cl_2$ (in the presence of $KOBu^t$) in CH_3CN and THF to afford luminescent complexes [1-(9-anthracylmethyl)-3-ethylimidazol-2-ylidene]$_2PdCl_2$ and [1-(9-anthracylmethyl)-3-ethylimidazol-2-ylidene]$_2$ $PtCl_2$, respectively [47] An analogous treatment of **84** with iodine afforded [1-(9-anthracylmethyl)-3-ethylimidazol-2-ylidene I]$^+$ I_3^-. In a refluxed CH_2Cl_2 solution imidazolium salt **84** is transformed into1-(9-anthracylmethyl)-3-ethylimidazol-2-ylidene chloride.

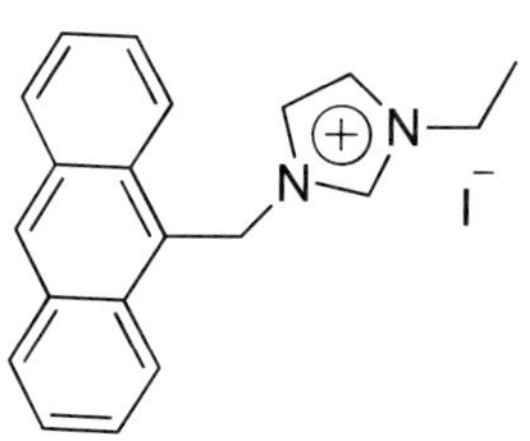

84

The complexes of Ag(I) and Au(I) (**85-86**) containing NHCs derived from cyclophanes were synthesized [48]. These complexes are reported to show higher cytotoxicity than cisplatin against several cancer cell lines including those resistant to cisplatin. The Au(I) complexes display higher anticancer activity than the corresponding Ag(I) complexes. Moreover, all the complexes are less cytotoxic than cisplatin on the normal human liver cell line LO2. The intracellular distribution studies show that these complexes are selectively localized in the mitochondria, especially for the Au(I) complexes. Interestingly, although Ag(I) as well as Au(I) complexes can influence mitochondrial integrity, they induce cancer cell death through different mechanisms. The Ag(I) complexes mainly induce caspase- and reactive oxygen species (ROS)-independent of early apoptosis. However, the Au(I) complexes induce both early and late apoptosis through caspase- and ROS-dependent pathways.

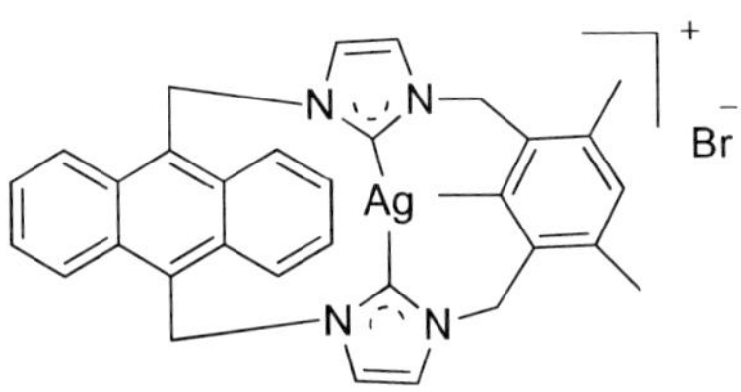

85

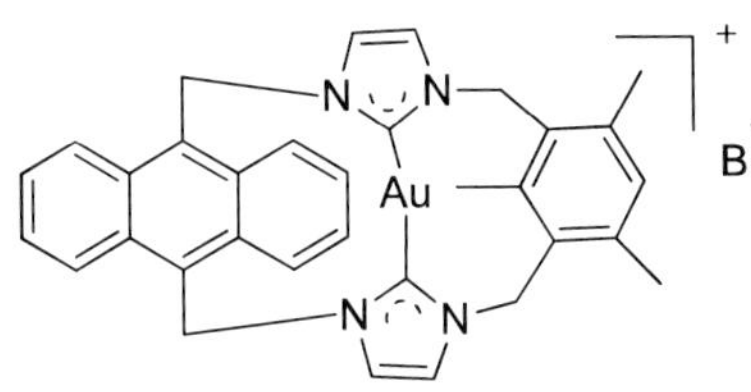

86

Mono and di-nuclear silver probes based on *N*-heterocyclic carbene having the anthracene chromophore (**87-88**) act as chemodosimeters for selective and sensitive detection of cyanide ions in the aqueous medium and signal the event through visible enhancement in fluorescence intensity [49]. The limit of detection of the cyanide ions was found to be 49 and 50 ppb, respectively, with **87** and **88**.

Half-sandwich ruthenium(II)-NHC complexes **89-90** [50] bearing anthracene were synthesized in good yield via transmetallation of the ligands followed by complexation with ruthenium precursor and characterized. The structure of the complex **90** was established with X-ray diffraction on its single crystal. The **89-90** exhibited catalytic activity for transfer hydrogenation of carbonyls in an alkaline medium using 2-propanol as a hydrogen source. The **89** showed the efficiency of ca. 100% at a catalyst loading of ca. 2 mol% within 2 h of the reaction in 2-propanol, whereas yield was quantitative in the presence **90** within 1 h at a low catalyst loading of ca. 0.5 mol%.

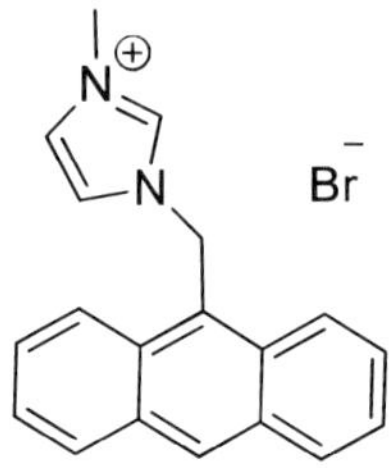

87

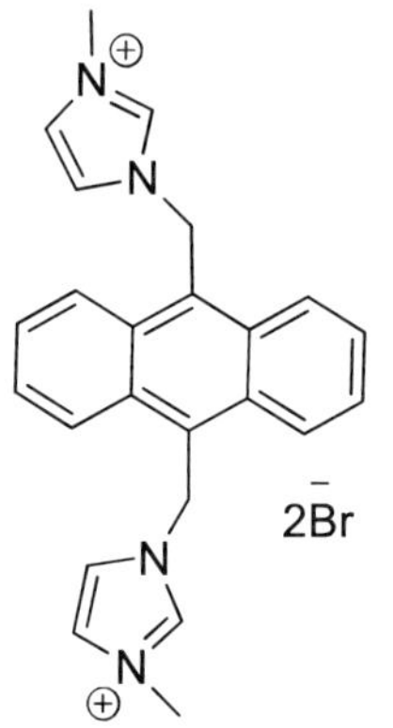

88

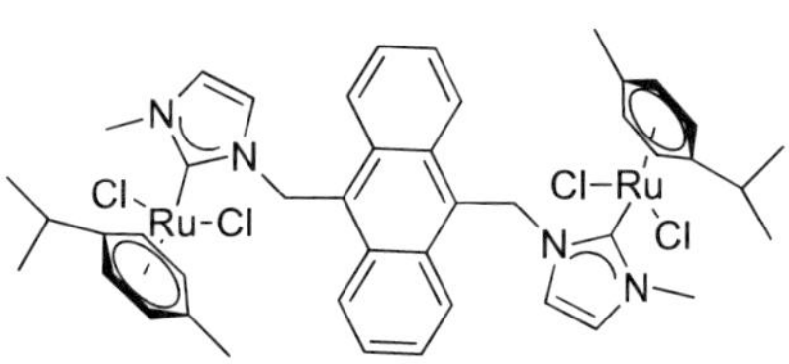

89

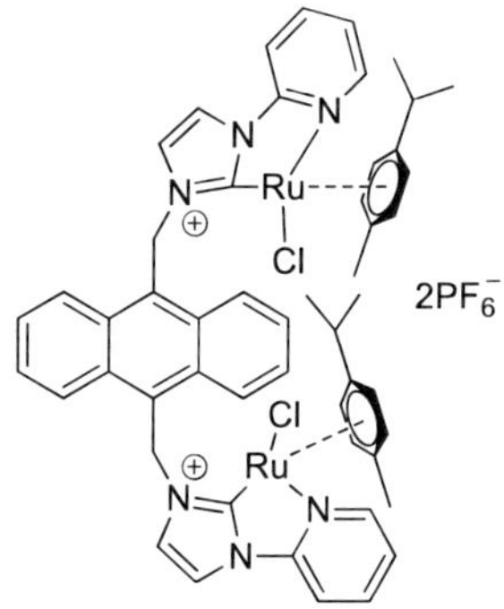

90

9,10 - Bis{di[2′ - (*N*-ethylimidazolium-1-yl)ethyl]aminomethyl} anthracenetetrakis(hexafluorophosphate) (**91**) was synthesized and characterized [51]. The catalytic activity of the **91**(0.2 - 0.5 mol% + $PdCl_2$ (1 mol%)) was tested in Suzuki - Miyaura, Heck - Mizoroki and Sonogashira reactions. The catalytic system was found effective for the three types of C - C coupling reactions.

N-Heterocyclic carbene palladium(II) complex containing two anthracene side arms (**92**) was immobilized on the surface of reduced graphene oxide (rGO) by π-stacking. The activity of **92** as such and the supported form in undirected C–H acetoxylation reaction of arenes was studied [52].

The results showed that for acetoxylation of benzene the catalytic efficiency with the immobilized complex was better compared to the homogeneous variant. According to XPS analysis, the immobilized catalyst maintains the original +II oxidation state of Pd after the catalytic reaction.

The *N*-heterocyclic carbene bi-silver(I) and bi-mercury(II) cyclophanes (**93-94**) in high yield were afforded. Their single crystal structures are reported. They have π- π stacking interactions [53].

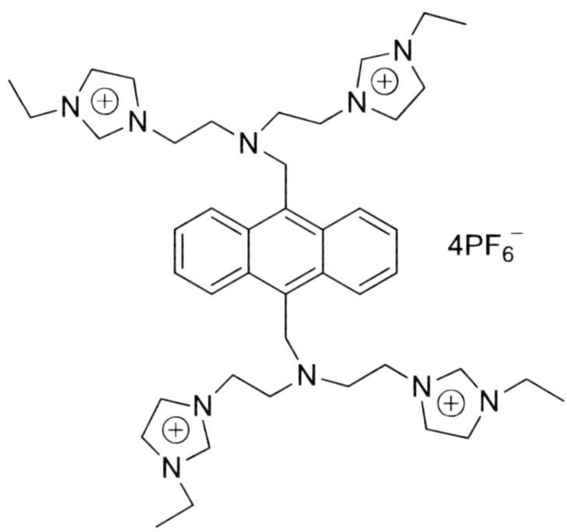

91

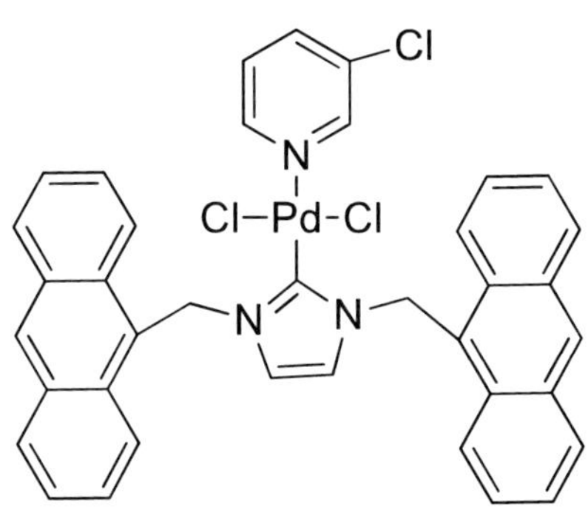

92

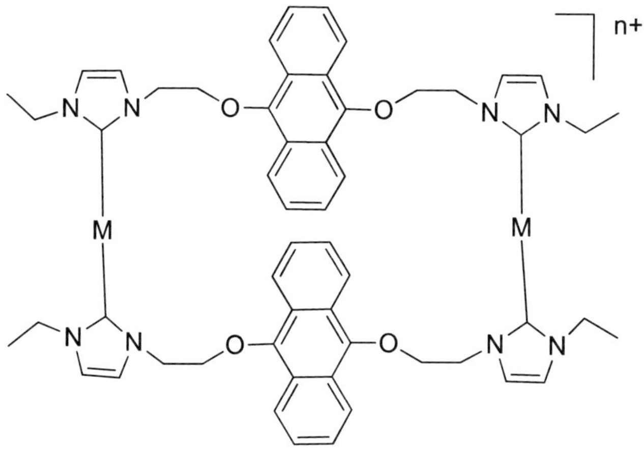

93; M = Ag n= 2
94; M = Ag n= 4

Chalcogen Bearing Ligands

The symmetric and asymmetric ligands based on a 1,8-disubstituted anthracene scaffold (**95-96**) were synthesized [54]. The ligands exhibit a nonplanar geometry in the solid state, owing to steric hindrance between the anthracene scaffolds. The **95** was complexed with [Mn(CO)$_5$Br] to afford the mononuclear species [(**95**)Mn(CO)$_3$Br] in which SMe groups chelated the Mn center in a *cis* fashion.

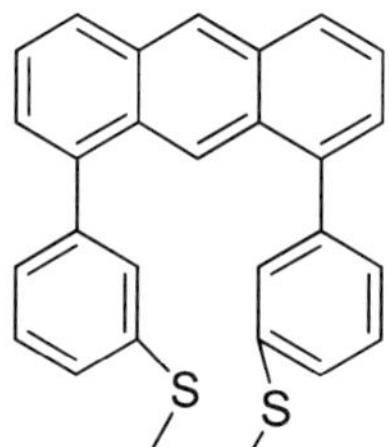

95

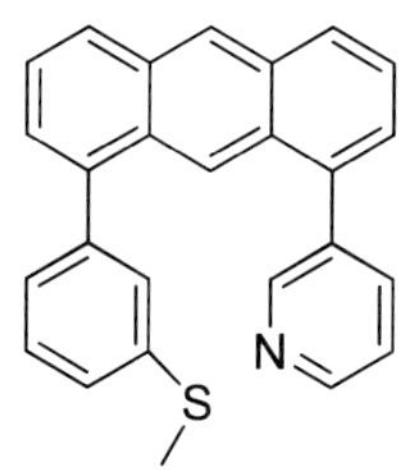

96

The **96** was found to coordinate through the py moieties, affording the bis-ligated complex [(**96**)$_2$Mn(CO)$_3$Br], wherein the thioether-S was unbound. The thiol analogue of **95** afforded the mixed-valence Mn^{I}/Mn^{II} dimer of formula [(Thiol analogue)$_3$Mn$_2$(CO)$_3$] having a *fac*-{Mn(CO)$_3$} unit supported by a triad of bridging thiolates, ligated to a supporting Mn(II) center. Three π-extended anthracene-bearing thioacetals (**97-99**) were synthesized, and their fluorescence "turn-on" responses to Hg^{2+} ions studied. The chemodosimetric fluorescence-sensing behavior and their resulting hydrolysis via a desulfurization reaction mechanism led to the formation of highly fluorescent respective aldehyde substitutions. The chemosensors **97-99** were found promising receptors toward Hg^{2+} ions in the presence of other competitive metal ions. The detection limits of **97-99** were found to be in the nanomolar range (94, 59, and 235, respectively). By fluorescence microscopic imaging studies **97-99** were found effective for fluorescence imaging in live cells [55].

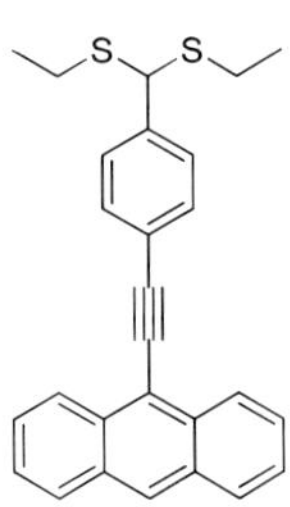

97

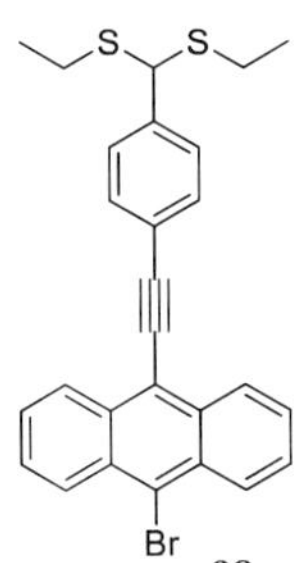

98

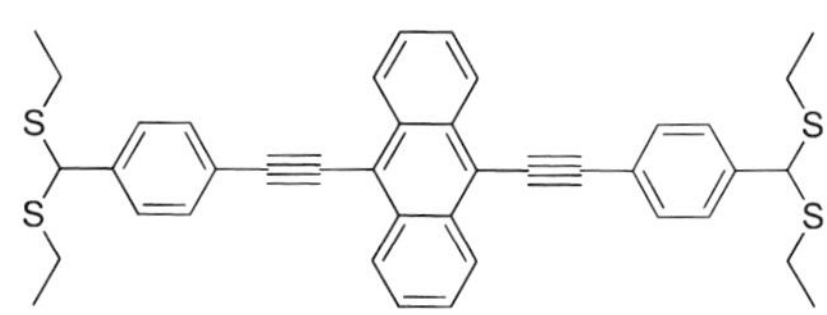

99

The condensation of anthracene-9-carbaldehyde with 2-(phenylthio/seleno)ethylamine resulted in Schiff bases [PhS(CH$_2$)$_2$CH=N-9-C$_{14}$H$_9$](**100**) and [PhSe(CH$_2$)$_2$CH=N-9-C$_{14}$H$_9$] (**101**). The iridacycles, [(η5-Cp*)Ir(**L**-H)][PF$_6$] (L = **100-101**) and complexes [(η^5-Cp*)Ir(**L**)Cl][PF$_6$] (**L** ligates in a bidentate mode) were synthesized and authenticated with HR-MS and NMR spectra [^{1}H, ^{13}C{^{1}H} and ^{77}Se{^{1}H}]. Single crystal structures of **101** and all half sandwich complexes are reported. The moisture and air insensitive complexes were found to act as efficient catalysts under mild conditions for base free *N*-alkylation of amines with benzyl alcohols and transfer hydrogenation (TH) of aldehydes/ketones [56].

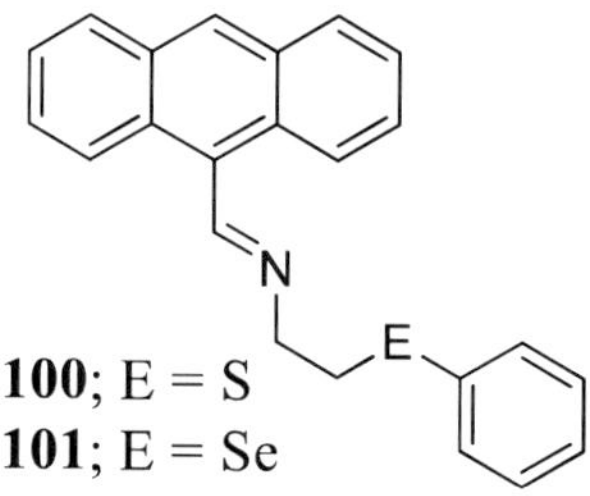

100; E = S
101; E = Se

The $Na_2[PdCl_4]$ on treatment with **100-101** has resulted in palladacycle [PdCl(C−, N, S/Se)] (**100/101** −H = (C−, N, S)/(C−, N, Se)). The reaction of $[(\eta^6\text{- }C_6H_6)RuCl(\mu\text{-}Cl)]_2$ with **100-101** in methanol for 8 h at room temperature (followed by addition of NH_4PF_6) afforded half-sandwich complex $[(\eta^6\text{-}C_6H_6)Ru(\mathbf{L})Cl][PF_6]$ (**L** = **100-101** as (N, E) ligand). The reaction of $[(\eta^5\text{-}Cp^*)RhCl(\mu\text{-}Cl)]_2$ with **100-101** in the presence of CH_3COONa at 50°C (followed by treatment with NH_4PF_6) resulted in $[(\eta^5\text{-}Cp^*)Rh($**100/101**-H)$][PF_6]$ On carrying out the reaction of $[(\eta^5\text{-}Cp^*)RhCl(\mu\text{-}Cl)]_2$ with these ligands at room temperature and in the absence of CH_3COONa, complex $[(\eta^5\text{-}Cp^*)Rh(\mathbf{L})Cl][PF_6]$, (**L** = **100-101** as (N, E) ligand) was formed. All complexes were authenticated with ^1H, $^{13}C\{^1H\}$, and $^{77}Se\{^1H\}$ NMR spectroscopy, high-resolution mass spectrometry, elemental analyses, and single-crystal X-ray diffraction. The moisture- and air-insensitive complexes of Pd(II) Ru(II) and Rh(III) were found thermally stable. Palladium and rhodium species efficiently catalyzed (under base-free condition) transfer hydrogenation (propan-2-ol as H-source). *N*-Alkylation of aniline with benzyl alcohol under base-free condition was promoted by Ru and Rh species. Mercury poisoning test supported homogeneous pathway for the two catalytic reactions. The rhodacycles probably gave real catalytic species by losing a Cp* group [57].

OXYGEN DONOR BEARING LIGANDS

The six complexes of Cu(II), Co(II), and Ni(II) with anthracene-9-carboxylic acid (**102**) were synthesized, incorporating auxiliary ligands

[2,2'-bipyridine, 1,4-diazabicyclo[2.2.2]octane, 1,10-phenanthroline, and 4,4'-bipyridine] and characterized. Their magnetic properties are reported. The complexes form different types of supramolecular assemblies using non bonding interactions [58]. The combination of CdSe semiconductor nanocrystals with 9-anthracene carboxylic acid ligands is reported to sensitize triplet−triplet annihilation on the emitter molecule diphenylanthracene. This hybrid system has shown to upconvert visible light (532 nm) to ultraviolet light (420 nm) [59]. The time-resolved photoluminescence measurements were used to characterize the kinetics of the energy transfer from the CdSe exciton state to the triplet state of the anthracene ligand. 9-Anthracene carboxylic acid binds to CdSe according to Poisson statistics with a maximum number of 2−3 ligands per nanocrystal. The CdSe-to-ligand energy transfer rate is $1.5 \times 107\ s^{-1}$. The overall efficiency of the energy transfer appears to be limited by the presence of fast non-radiative decay channels in the nanocrystals and the low coverage of anthracene ligands [60]. The Cd(II) complexes [61] with anthracene-9,10-dicarboxylate (**103**) and auxiliary ligands, 2,2'-bipyridine, 1,10-phenanthroline, 3-(2-pyridyl)pyrazole, and 1,5-diazacyclooctane were synthesized and characterized structurally including those of their single crystals. The complexes are reported to have luminescent properties.

The complexes of cobalt [Co(**104**)(TPA)]PF_6 and [Co(**104**)(Me_nTPA)]PF_6 (n = 1-3) were synthesized [62] by the combination of 9-(3,4-dihydroxyphenyl)anthracene (H_2**104**) and tris-(2-pyridylmethyl)amine (TPA) or its derivatives (Me_nTPA, n = 1, 2, 3). The characterization of complexes was performed with UV-vis, IR, ^{1}H NMR, and magnetic susceptibility measurements. In the solid state, the variable-temperature magnetic susceptibility data showed that the first complex is low-spin cobalt(III) catecholate (CoIII(LS)-Cat), while complex for n= 3 is high-spin cobalt(II) semiquinonate (CoII(HS)-SQ) in the temperature range 4.5–400 K.

The susceptibility data of the complexes, n = 2 and 3 suggested valence tautomerism between the CoIII(LS)-Cat and CoII(HS)-SQ forms. Light-induced valence tautomerism was observed in the complexes of n = 2 and 3 at 5 K by photo-irradiation.

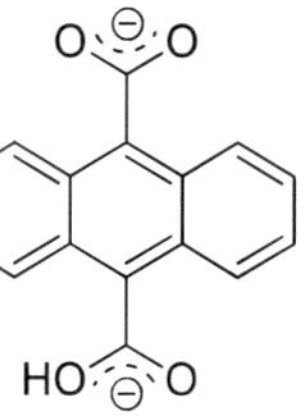

103

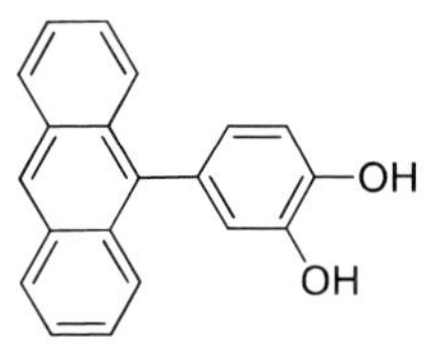

104

The anthracene-9-carboxylic acid (**102**) as a primary ligand, and quinoline-2-carboxylic acid as the second ligand were used to construct a tin(IV)-carboxylate coordination complex [63] under thermal gradient condition. The complex was characterized by ^{1}H and $^{13}C\{^{1}H\}$ NMR, IR and UV spectra and elemental analyses. The molecular structure of the complex was determined by single-crystal X-ray diffraction. The nanocrystalline form of the complex was successfully obtained at 30, 50 and 60°C by a facile sonochemical route. The nanocrystalline complexes were characterized by scanning electron microscopy, X-ray powder diffraction, IR spectroscopy and elemental analyses. The photoluminescence properties of the nanocrystalline and crystalline bulk complex in solid-state indicated a remarkable effect of the size of the complex particles on its optical properties. The absorption and emission peaks of the nanocrystalline complex were found blue shifted significantly in comparison to those of the single-crystal form. The application of the prepared complex in the fabrication of an organic light-emitting diode was demonstrated. The current–voltage (I–V) characteristics and the electroluminescence (EL) properties of the complex have been investigated. The EL of the compound exhibits green emission at 552 nm.

The orange colored cubic-shaped single crystals of Zr-MOF with a formula of $[Zr_6O_4(OH)_4(\mathbf{L})_6]\cdot 6DMF$ were prepared by a solvothermal reaction of $ZrCl_4$ with 4,4'-(anthracene-9,10-diylbis(ethyne-2,1-diyl))dibenzoic acid ($H_2\mathbf{L}$) and used as a visible-light photocatalyst for carbon dioxide reduction to formate. The MOF was found highly efficient for visible-light-driven CO_2 reduction as formation rate of the formate was found to be 183.3 mmol h^{-1} mmol $(MOF)^{-1}$. Both the inorganic building unit Zr_6 oxo cluster and the anthracene-based ligand contribute to the photocatalysis of CO_2 reduction. The dual photocatalytic route was demonstrated to be more efficient for visible-light-driven CO_2 photoreduction than the ligand-to metal charge transfer [64]. The mixture of 4,4'-(anthracene-9,10-diylbis(ethyne-2,1-diyl))-dibenzoic acid) and $Zn(NO_3)_2{\cdot}6H_2O$, $NaNO_3$ and *N,N*-diethylformamide was placed in a capped glass vial at 85°C for 72 h to afford the needle-like crystals of a MOF [65a]. The rarely occurring long-range π-stacking of the ligand in the form of a zig-zag chain was found in the MOF structure, which provided not only a new charge transport pathway with high electrical conductivity of 1.3 (±0.5) x 10^{-3} S cm^{-1} but also an electroluminescence property with an emission centred at 575 nm. The zirconium(IV) complexes of ligand constructed on anthracene (**104a**) were synthesized. They were found competent for the catalytic formation of pyrimidines from alkynes and nitriles [65b].

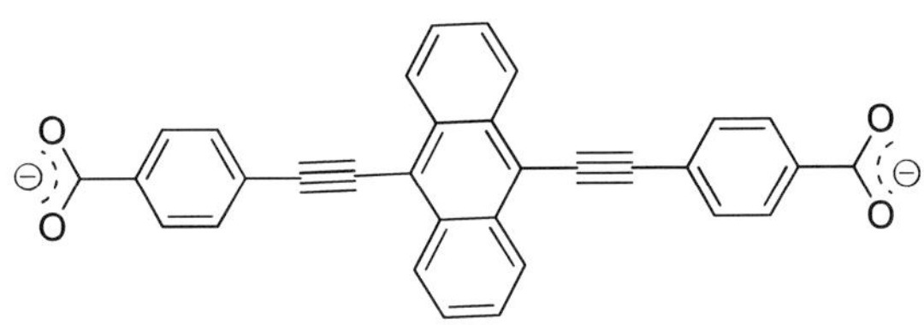

104a

ORGANOMETALLIC COMPOUNDS

The homodinuclear iron complex [Cp*Fe(μ-η^4:η4-L)FeCp*] (L = $C_{14}H_{10}$) was synthesized and characterized by ^1H and $^{13}C\{^1H\}$ NMR, UV/Vis, and ^{57}Fe Mössbauer spectroscopy [66]. Its molecular structure was determined by X-ray crystallography. The complex was found diamagnetic as a result of strong magnetic coupling of its17e Fe(I) centers mediated by polyarene bridge. The analysis of its redox behavior by cyclic voltammetry and UV/Vis spectroelectrochemistry has shown that it can be oxidized reversibly in two well-separated one-electron steps to monocation $[Cp^*Fe(\mu L)FeCp^*]^+$ and dication $[Cp^*Fe(\mu\text{-}L)FeCp^*]^{2+}$. The reduction to a monoanion $[Cp^*Fe(\mu\text{-}L)FeCp^*]^-$ has been also observed.

The four anthracene-based bimetallic ruthenium vinyl complexes [67], in which the ruthenium units were attached at different positions (9,10, 1,5, 2,6, and 1,8) of the anthracene moiety, were synthesized by treating the appropriate anthracene-based ethynes with $[RuHCl(CO)(PPh_3)_3]$. The bimetallic complexes were characterized by NMR, X-ray diffraction, and elemental analyses. The single-crystal X-ray structures of the complexes revealed more planar structure for the 2,6-disubstituted ruthenium vinyl complex compared to the 9,10-disubstituted one, possibly because of less steric hindrance. The 2,6-disubstituted ruthenium vinyl complex displayed the strongest fluorescence emission due to the planarity of its organic conjugated bridge. The electrochemical studies showed that the two ruthenium centers displayed obvious differences in electronic communication when located at different positions on the anthracene unit. The 9,10- and 1,8-disubstituted ruthenium vinyl complexes exhibited better electronic communication and higher stability of the mixed-valence complex than the 1,5- and 2,6-disubstituted analogues.

The 1,8-dichloro-10-(trimethylsilyl)-, 1,8-dichloro-10-(trimethylgermyl)- and 1,8-dichloro-10-(trimethylstannyl)anthracene were synthesized and characterized by multinuclear NMR spectroscopy and mass spectrometry. There is Si/Ge /Sn-C σ bond in all these compounds [68]. Their molecular structures in the crystalline state were analyzed with X-ray diffraction and compared with the crystal structure of 10-*tert*-butyl-1,8-dichloroanthracene. It was found that the level of deformation of the anthracene backbone follows the order $SnMe_3 < GeMe_3 < SiMe_3 < CMe_3$.

Platinum acetylide oligomers (**105**) that contain an anthracene moiety were synthesized and characterized [69]. The spectroscopic measurements and DFT calculations revealed that both the singlet and triplet energy levels of the anthracene segment were lower than those of the platinum acetylide segment. Thus, the platinum acetylide segment acted as a sensitizer to populate the triplet state of the anthrancene segment via intramolecular triplet–triplet energy transfer. The fluorescence quenching and the dominant triplet absorption that arose from the anthracene segment in the transient absorption spectrum of **105** gave clear evidence that energy transfer adopted an indirect mechanism, which began with singlet–triplet energy transfer from the anthracene segment to the platinum acetylide segment followed by triplet−triplet energy transfer to the anthracene segment.

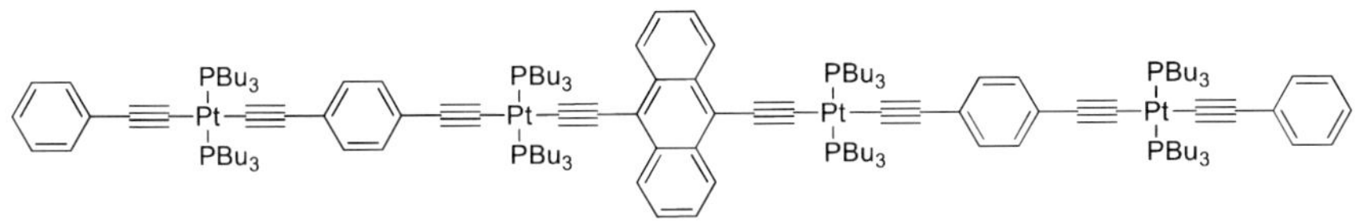

105

Homobimetallic anthracene-bridged η^5-cyclopentadienyl derivatives of rhodium(I) (**106**) were reported [70]. They were found poorly luminescent compounds which showed highly resolved EPR spectra indicating such species as average-valence [$Rh^{+1/2}$, $Rh^{+1/2}$] complexes.

REFERENCES

[1] a) Becker, H. D. (1993). "Unimolecular Photochemistry of Anthracenes." *Chemical Reviews*, *93*,145–172, b) Bouas-Laurant, H., Castellan, A., Desvergne, J. P. & Laypouyade, R. (2000). "Photodimerization of Anthracenes in Fluid Solution: Structural Aspects" *Chemical Society Review*, *29*, 43–55, c) Bouas-Laurant, H., Castellan, A., Desvergne, J. P. & Laypouyade, R. (2001). "Photodimerization of Anthracenes In Fluid Solutions: (Part 2) Mechanistic Aspects of The Photocycloaddition and of The Photochemical and Thermal Cleavage." *Chemical Society Review*, *30*, 248–263.

[2] a) Chmiel, J., Neumann, B., Stammler, H. G. & Mitzel, N. W. (2010). "Dialkylaluminium-, -Gallium-, and -Indium-Based Poly-Lewis Acids with a 1,8-Diethynylanthracene Backbone." *Chemistry A European Journal*, *16*, 11906–11914, b) Katz, H. E. (1989). "1,8-Anthracenediethynylbis(catechol boronate): A Bidentate Lewis Acid on a Novel Framework." *Journal of Orgranic Chemistry*, *54*, 2179–2183.

[3] Kraicheva, I., Tsacheva, I., Vodenicharova, E., Tashev, E., Tosheva, T., Kril, A., Topashka-Ancheva, M., Iliev, I., Gerasimova, Ts. & Troev, K. (2012). "Synthesis, Antiproliferative Activity and Genotoxicity of Novel Anthracene-Containing Aminophosphonates and A New Anthracene-Derived Schiff Base." *Bioorganic & Medical Chemistry*, *20*, 117–124.

[4] Erdemir, S. & Malkondu, S. (2015). "Novel "Turn On" Fluorescent Sensors Based on Anthracene and Carbazone Units for Cu(II) Ion in $CH_3CN–H_2O$." *Journal of Luminescence*, *158*, 86–90.

[5] Andiappan, K., Sanmugam, A., Deivanayagama, E., Karuppasamy, K., Kim, H. S. & Vikraman, D. (2019). "Schiff Base Rare Earth Metal Complexes: Studies on Functional, Optical and Thermal Properties and Assessment of Antibacterial Activity." *International Journal of Biological Macromolecules*, *124*, 403–410.

[6] Punithakumari, G. & Velmathi, S. (2019). "Smart Sensing of Cyanide and Iron(III) by Anthracene Based Probe Through Relay Recognition Approach." *Journal of Photochemistry Photobiological A: Chemistry*, *373*, 94–104.

[7] Hwang, S. M., Kim, A. & Kim, C. (2019). "A Simple Hydrazine-Based Probe Bearing Anthracene Moiety for The Highly Selective Detection of Hypochlorite." *Inorganic Chemistry Communications*, *101*, 1–5.

[8] Crespo, M. & Evangelio, E. (2004). "Five- and Six-Membered Platinacycles Derived From Phenantryl and Anthracenyl Imines" *Journal of Organometallic Chemistry*, *689*, 1956–1964.

[9] Zhou, J., Wang, Q. & Sun, H. (2008). "Di-L-Chlorido-Bis({9-[(2,6-Diisopropylphenyl)Iminomethyl]Anthracen-1-Yl}- Palladium(II))." *Acta Crystallographica*, *E64*, m688.

[10] Zhou, J., Li, X. & Sun, H. (2010). "An Efficient and Recyclable Water-Soluble Cyclopalladated Complex for Aqueous Suzuki Reactions Under Aerial Conditions." *Journal of Organometallic Chemistry*, *695*, 297–303.

[11] Rao, A. R. B. & Pal, S. (2012). "Cyclopalladated Complexes with 9-Anthraldehyde Aroylhydrazones: Synthesis, Properties and Structures." *Journal of Organometallic Chemistry*, *701*, 62–67.

[12] Kate, A. N., Kumbhar, A. A., Khan, A. A., Joshi, P. V. & Puranik, V. G. (2014). "Monitoring Cellular Uptake and Cytotoxicity of Copper(II) Complex Using A Fluorescent Anthracene Thiosemicarbazone Ligand" *Bioconjugate Chemistry*, *25*, 102−114.

[13] Bai, J., Wang, R.-H., Qiao, Y., Wang, A., Fang, C.-J. 2017. "Schiff Base Derived from Thiosemicarbazone and Anthracene Showed High Potential in Overcoming Multi Drug Resistance in Vitro with Low Drug Resistance Index." *Drug Design, Development Therapy*, *11*, 2227-2237.

[14] Shellaiah, M., Wu, Y. H., Singh, A., Raju, M. V. R. & Lin, H. C. (2013). "Novel Pyrene- and Anthracene-Based Schiff Base Derivatives as Cu^{2+} And Fe^{3+} Fluorescence Turn-On Sensors and for Aggregation Induced Emissions." *Journal of Materials Chemistry A*, *1*, 1310–1318.

[15] Yang, M., Zhang, Y., Zhu, W., Wang, H., Huang, J., Cheng, L., Zhou, H., Wua, J. & Tiana, Y. (2015). "Difunctional Chemosensor for Cu(II) And Zn(II) Based on Schiff Base Modified Anthryl Derivative with Aggregation-Induced Emission Enhancement and Piezochromic Characteristics." *Journal of Materials Chemistry C*, *3*, 1994–2002.

[16] Jaividhya, P., Ganeshpandian, M., Dhivy, R., Akbarsha, M. A. & Palaniandavar, M. (2015). Fluorescent Mixed Ligand Copper(II) Complexes of Anthracene-Appended Schiff Bases: Studies on DNA Binding, Nuclease Activity and Cytotoxicity." *Dalton Transaction*, *44*, 11997–12010.

[17] Densil, S., Chang, C. H., Chen, C. L., Mathavan, A., Ramdass, A., Sathish, V., Thanasekaran, P., Li, W. S. & Rajagopal, S. (2018). "Aggregation-Induced Emission Enhancement of Anthracene-derived Schiff Base Compounds and Their Application as a Sensor for Bovine Serum Albumin and Optical Cell Imaging." *Luminescence*, *33*, 780–789.

[18] Gubendran, A., Gangatharan, G., Kumar, V., Kesavan, M. P., Rajagopal, G., Athappan, P. & Rajesh, J. (2018). "New Anthracene Based Schiff Base Ligands Appended Cu(II) Complexes, Theoretical Study, DNA Binding and Cleavage Activities." *Applied Organometallic Chemistry*, *32*, e4128.

[19] Kathiresan, S., Annaraj, J. & Bhuvanesh, N. S. P. (2017). "Cu(II) and Ni(II) Complexes of Anthracene-Affixed Schiff Base: A Conflict between Covalent and Stacking Interactions with DNA Bases." *Chemistry Select*, *2*, 5475–5484.

[20] Liu, S., Liu, L., Liang, C., Peng, Y., Huang, S., Li, X., Sun, M. & Wang, X. (2018). "Non-Conjugated Anthracene Derivatives and their Mechanofluorochromic Properties." *Research on Chemical Intermediates*, *44*, 3923–3931.

[21] Htwe, T., Muppidi, V. K., Pradeep, C. P., Zacharias, P. S. & Pal, S. (2006). "Nickel(II) Complexes of Tridentate Anthracene Based Schiff Bases: Syntheses, Properties and Crystal Structures." *Journal of Coordination Chemistry*, *59*, 671–680.

[22] Andiappan, K., Sanmugam, A., Deivanayagam, E., Karuppasamy, K., Kim, H. S. & Vikraman, D. (2018). "*In Vitro* Cytotoxicity Activity of Novel Schiff Base Ligand–Lanthanide Complexes *Scientific Report*, *8*, 3054.

[23] Salimon, J. & Salih, N. (2010). "New Schiff Bases Derivatives Containing Anthracene and 1,3,4-Thidiazole Moieties: Synthesis and Fungicidal Activity." *International Journal of Pharm Tech Research*, *2*, 205–208.

[24] a) Wang, W., Yua, X., Zhanga, Q., Lia, M., Jianga, X., Liua, Q. & Gaob, Y. (2014). "Synthesis and Fe^{3+} Recognition of Novel Triazole Schiff-Base Derivatives Bearing Fluorogenic Naphthalene and Anthracene Units." *Journal of Chemical Research*, *38*, 627–629, b) Simon, T., Shellaiah, M., Srinivasadesikan, V., Lin, C. C., Ko, F. H., Sunb, K. W. & Linc, M. C. (2016). "Novel Anthracene- and Pyridine-Containing Schiff Base Probe for Selective "Off–On" Fluorescent Determination of Cu^{2+} Ions Towards Live Cell Application." *New Journal of Chemistry*, *40*, 6101–6108.

[25] Yu, M. X., Duan, J. P., Lin, C. H., Cheng, C. H. & Tao, Y. T. (2002). "Diaminoanthracene Derivatives as High-Performance Green Host Electroluminescent Materials." *Chemistry of Materials*, *14*, 3958–3963.

[26] a) Noto, N., Tanaka, Y., Koike, T. & Akita, M. (2018). "Strongly Reducing (Diarylamino)anthracene Catalyst for Metal-Free Visible-Light Photocatalytic Fluoroalkylation." *ACS Catalysis*, *8*, 9408–9419, b) Suktanarak, P., Watchasit, S., Chitchak, K., Plainpan, N., Chainok, K., Vanalabhpatana, P., Pienpinijtham, P., Suksai, C., Tuntulani, T., Ruangpornvisuti, V. & Leeladee, P. (2018). "Tuning The Reactivity of Copper Complexes Supported by Tridentate Ligands Leading to Two-Electron Reduction of Dioxygen." *Dalton Transactions*, *47*, 16337–16349.

[27] Jia, L. M., Tong, J. & Yu, S. Y. (2018). "Neutral Coordination Polymers of Cobalt(II), Copper(II), Zinc(II) and Manganese(II) B-Diketonate Complexes with Fluorescent Anthracene Dipyridine: Synthesis, Structure and Luminescence Properties." *Journal of Photochemistry and Photobiology A: Chemistry*, *355*, 84–93.

[28] Yao, L. Y., Qin, L., Xie, T. Z., Li, Y. Z. & Yu, S. Y. (2011). "Synthesis and Anion Sensing of Water-Soluble Metallomacrocycles." *Inorganic Chemistry*, *50*, 6055–6062.

[29] Kuehl, C. J., Mayne, C. L., Arif, A. M. & Stang, Peter J. (2000). "Coordination-Driven Assembly of Molecular Rectangles via an Organometallic "Clip"." *Org. Lett.*, *2*, 3727–3729.

[30] Saranya, G., Yam, C., Gao, S. & Chen, M. (2018). "Roles of Chenodeoxycholic Acid Coadsorbent in Anthracene-Based Dye-Sensitized Solar Cells: A Density Functional Theory Study." *Journal of Physical Chemistry C*, *122*, 23280−23287.

[31] Nair, A. K., Neelakandan, P. P. & Ramaiah, D. (2009). "A Supramolecular Cu(II) Metallocyclophane Probe for Guanosine 5'-Monophosphate." *Chemical Communication*, 6352–6354.

[32] Ramaiah, D., Neelakandan, P. P., Nair, A. K. & Avirah, R. R. (2010). "Functional Cyclophanes, Promising Hosts for Optical Biomolecular Recognition." *Chemical Society Review*, *39*, 4158–4168.

[33] Askarizadeh, E., Devoille, A. M., J. Boghaei, D. M., Slawin, A. M. Z. & Love, J. B. (2009). "Ligand Modifications for Tailoring the Binuclear Microenvironments in Schiff-Base Calixpyrrole Pacman Complexes." *Inorganic Chemistry*, *48*, 7491–7500.

[34] Devoille, A. M. J. & Love, J. B. (2012). "Double-Pillared Cobalt Pacman Complexes: Synthesis, Structures and Oxygen Reduction Catalysis." *Dalton Transactions*, *41*, 65–72.

[35] Ganeshpandian, M., Loganathan, R., Suresh, E., Riyasdeen, A., Akbarshad, M. A. & Palaniandavar, M. (2014). "New Ruthenium(II) Arene Complexes of Anthracenylappended Diazacycloalkanes: Effect of Ligand Intercalation and Hydrophobicity on DNA and Protein Binding and Cleavage and Cytotoxicity." *Dalton Transactions*, *43*, 1203–1219.

[36] Liu, W., Ying, S., Guo, R., Qiao, X., Leng, P., Zhang, Q., Wang, Y., Ma, D. & Wang, L. (2019). "Nondoped Blue Fluorescent Organic Light-Emitting Diodes Based on Benzonitrile-Anthracene Derivative with 10.06% External Quantum Efficiency and Low Efficiency Roll-Off." *Journal of Materials Chemistry C*, *7*, 1014–1021.

[37] Sarkar, S., McGowan, K. P., Culver, J. A., Carlson, A. R., Koller, J., Peloquin, A. J., Veige, M. K., Abboud, K. A. & Veige, A. S. (2010). "NCN Trianionic Pincer Ligand Precursors: Synthesis of Bimetallic, Chelating Diamide, and Pincer Group IV Complexes." *Inorganic Chemistry*, *49*, 5143–5156.

[38] Tao, Q., Yan, D., Liao, Y., Huang, X., Deng, J. & Yu, D. (2019). "Synthesis and Photovoltaic Performance of Anthracene-Based Small Molecules for Solution-Processed Organic Solar Cells." *Chemistry Select*, *4*, 752–758.

[39] a) Haenel, M., Jakubik, W., Krüger, D. C. & Betz, P. (1991). "1,8-Bis(diphenylphosphino)anthracene and Metal Complexes." *Chemische Berichte*, *124*, 333–336, (b) Cao, D. K., Gu, Y. W., Feng, J. Q., Caia, Z. Sh. & Ward, M. D. (2013). "Mononuclear Lanthanide Complexes Incorporating an Anthracene Group: Structural Modification, Slow Magnetic Relaxation and Multicomponent

Fluorescence Emissions in Dy Compounds." *Dalton Transactions*, *42*, 11436–11444.

[40] Azerraf, C., Shpruhman, A. & Gelman, D. (2009). "Diels–Alder Cycloaddition as a New Approach Toward Stable PC(Sp^3)P-Metalated Compounds." *Chemical Communication*, 466–468.

[41] Wong, W. Y., Ting, F. L., Guo, Y. Z. & Lin. (2005). "Synthesis, Characterization and Crystal Structures of New Ruthenium Carbonyl Clusters Derived from (9-Anthracenyl)diphenylphosphine, *J. Cluster Sci.*, *16*, 185-200.

[42] Kim, S. K., Singh, N. J., Kwon, J., Hwang, I.-C., Park, S. J., Kimb, K. S. & Yoon, J. (2006). "Fluorescent Imidazolium Receptors for the Recognition of Pyrophosphate." *Tetrahedron*, *62*, 6065–6072.

[43] a) Qin, D. B., Xu, F. B., Wan, X. J., Zhao, Y. J. & Zhang, Z. Z. (2006). "Effectively Selective Fluorescent Chemosensor for Terephthalate." *Tetrahedron Letters*, *47*, 5641–5643, b) Ho, J. H. H., Choy, S. W. S., Macgregor, S. A. & Messerle, B. A. (2011). "Cooperativity in Bimetallic Dihydroalkoxylation Catalysts Built on Aromatic Scaffolds: Significant Rate Enhancements with a Rigid Anthracene Scaffold." *Organometallics*, *30*, 5978−5984.

[44] Ghosh, K., Saha, I. & Patra, A. (2009). "Design and Synthesis of an Ortho-Phenylenediamine-Based Open Cleft: A Selective Fluorescent Chemosensor for Dihydrogen Phosphate." *Tetrahedron Letters*, *50*, 2392–2397.

[45] Xu, Z., Kim, S., Lee, K. H. & Yoon, J. (2007). "A Highly Selective Fluorescent Chemosensor for Dihydrogen Phosphate via Unique Excimer Formation and PET Mechanism." *Tetrahedron Letters*, *48*, 3797–3800.

[46] Ghosh, K. & Saha, I. (2008). "Anthracene-Based Ortho-Phenylenediamine Clefts for Sensing Carboxylates." *Tetrahedron Letters*, *49*, 4591–4595.

[47] Liu, Q. X., Song, H. B., Xu, F. B., Li, Q. S., Zeng, X. S., Leng, X. B. & Zhang, Z. Z. (2003). "Synthesis, Crystal Structure and Photophysical Properties of *N*-Heterocyclic Carbene Pd(II), Pt(II) Complexes and Iodine Adduct." *Polyhedron*, *22*, 1515–1521.

[48] Li, Y., Liu, G. F., Tan, C. P., Ji, L. N. & Mao, Z. W. (2014). "Antitumor Properties and Mechanisms of Mitochondria-Targeted Ag(I) And Au(I) Complexes Containing N-Heterocyclic Carbenes Derived from Cyclophanes." *Metallomics*, *6*, 1460–1468.

[49] Viji, M., Nair, A. K., Nandajana, P. C. & Ramaiah, D. (2014). "Fluorescent Chemodosimeter Based On NHC Complex for Selective Recognition of Cyanide Ions in Aqueous Medium." *RSC Advances*, *4*, 47982–47986.

[50] Viji, M., Tyagi, N., Naithani, N. & Ramaiah, D. (2017). "Aryl Appended Neutral and Cationic Half-Sandwich Ruthenium(II)–NHC Complexes: Synthesis, Characterisation and Catalytic Applications." *New Journal of Chemistry*, *41*, 12736–12745.

[51] Liu, Q. X., Zhao, D. X., Wu, H., Zhao, Z. X. & Lv, S. Z. (2018). Catalytic Activities of NHC-$PdCl_2$ Species Based On Functionalized Tetradentate Imidazolium Salt In Three Types of C-C Coupling Reactions." *Applied Organometallic Chemistry*, *32*, e4429.

[52] Majeed, M. H., Shayesteh, P., Persson, A. R., Wallenberg, L. R., Schnadt, J. & Wendt, Ola F. (2018). "A Pd^{II} Carbene Complex with Anthracene Side-Arms for π-Stacking on Reduced Graphene Oxide (rGO): Activity towards Undirected C–H Oxygenation of Arenes." *European Journal of Inorganic Chemistry*, 4742–4746.

[53] Wan, X. J., Xu, F. B., Li, Q. S., Song, H. B. & Zhang, Z. Z. (2005). "Synthesis and Crystal Structures of Binuclear Metallocyclophane with N-Heterocyclic Carbene Linkage by the *π-π* Stacking Interaction." *Organometallics*, *24*, 6066-6068.

[54] Manes, T. A. & Rose, M. J. (2016). "Mono- and Dinuclear Manganese Carbonyls Supported by 1,8-Disubstituted (L = Py, SMe, SH) Anthracene Ligand Scaffolds." *Inorganic Chemistry*, 55, 5127−5138.

[55] Swamy, P. C. A., Shanmugapriya, J., Singaravadivel, S., Sivaraman, G. & Chellappa, D. (2018). "Anthracene-Based Highly Selective and Sensitive Fluorescent "Turnon" Chemodosimeter for Hg^{2+}." *ACS Omega*, *3*, 12341−12348.

[56] Dubey, P., Gupta, S. & Singh, A. K. (2018). "Base Free N-Alkylation of Anilines with $ArCH_2OH$ and Transfer Hydrogenation of Aldehydes/Ketones Catalyzed By The Complexes of η^5-Cp*Ir(III) with Chalcogenated Schiff Bases of Anthracene-9-carbaldehyde." *Dalton Transactions*, *47*, 3764–3774.

[57] Dubey, P., Gupta, S. & Singh, A. K. (2019). "Complexes of Pd(II), η6 -C_6H_6Ru(II), and η5 -Cp*Rh(III) with Chalcogenated Schiff Bases of Anthracene-9-carbaldehyde and Base-Free Catalytic Transfer Hydrogenation of Aldehydes/Ketones and N-Alkylation of Amines." *Organometallics*, *38*, 944–961.

[58] Liu, C. S., Wang, J. J., Yan, L. F., Chang, Z., Bu, X. H., Sañudo, E. C. & Ribas, J. (2007). "Copper(II), Cobalt(II), and Nickel(II) Complexes with a Bulky Anthracene-Based Carboxylic Ligand, Syntheses, Crystal Structures, and Magnetic Properties." *Inorganic Chemistry*, *46*, 6299−6310.

[59] Huang, Z., Li, X., Mahboub, M., Hanson, K. M., Nichols, V. M., Le, H., Tang, M. L. & Bardeen, C. J. (2015). "Hybrid Molecule−Nanocrystal Photon Upconversion Across the Visible and Near-Infrared." *Nano Letters*, *15*, 5552−5557.

[60] Piland, G. B., Huang, Z., Tang, M. L. & Bardeen, C. J. (2016). "Dynamics of Energy Transfer from CdSe Nanocrystals to Triplet States of Anthracene Ligand Molecules." *Journal of Physical Chemistry C*, *120*, 5883−5889.

[61] Liu, C. S., Wang, J. J., Chang, Z., Yana, L. F. & Bu, X. He. (2010). "Cadmium(II) Coordination Polymers Based on A Bulky Anthracene-Based Dicarboxylate Ligand: Crystal Structures and Luminescent Properties." *Cryst Eng Comm*, *12*, 1833–1841.

[62] Katayama, K., Hirotsu, M., Kinoshita, I. & Teki, Y. (2014). "Synthesis, Magnetic Properties and Dynamic Behavior of Cobalt Complexes with An Anthracenecontaining Dioxolene Ligand." *Dalton Transactions*, *43*, 13384–13391.

[63] Hodaie, M., Sadjadi, M. S., Amini, M., Najafi, M. E. S. & Ng, W. (2016). "Sonochemical Synthesis of a Nano crystalline Tin(IV) Complex based on a Bulky Anthracene Carboxylate Ligand, Spectroscopic and Photophysical Properties" *J. Inorg. Organomet. Polym.*, *26*, 500–511.

[64] Chen, Da., Xing, H., Wang, C. & Su, Z. (2016). "Highly Efficient Visible-Light-Driven CO_2 Reduction to Formate by A New Anthracene-Based Zirconium MOF via Dual Catalytic Routes." *Journal of Materials Chemistry A*, *4*, 2657–2662.

[65] a) Chen, D., Xing, H., Su, Z. & Wang, C. (2016). "Electrical Conductivity and Electroluminescence of A New Anthracene-Based Metal–Organic Framework with π-Conjugated Zigzag Chains." *Chemical Communication*, *52*, 2019–2022, b) Low, C. H., Rosenberg, J. N., Lopez, M. A. & Agapie, T. (2018). "Oxidative Coupling with Zr(IV) Supported by a Noninnocent Anthracene-Based Ligand: Application to the Catalytic Cotrimerization of Alkynes and Nitriles to Pyrimidines." *Journal of the American Chemical Society*, *140*, 11906−11910.

[66] Schnöckelborg, E. M., Hartl, F., Langer, T., Pöttgen, R. & Wolf, R. (2012). "Redox-Active, Dinuclear Sandwich Compounds [Cp*Fe(μ-L)FeCp*](L = Naphthalene and Anthracene)." *European Journal of Inorganic Chemistry*, 1632–1638.

[67] Ou, Y. P., Jiang, C., Wu, D., Xia, J., Yin, J., Jin, S., Yu, G. A. & Liu, S. H. (2011). "Synthesis, Characterization, and Properties of Anthracene-Bridged Bimetallic Ruthenium Vinyl Complexes $[RuCl(CO)(PMe_3)_3]_2$(μ-CH=CH-anthracene-CH=CH)." *Organometallics*, *30*, 5763–5770.

[68] Lamm, J. H., Vishnevskiy, Y. V., Ziemann, E., Kinder, T. A., Neumann, B., Stammler, H. G. & Mitzel, N. W. (2014). "Syntheses and Structures of 10-Trimethylelement-Substituted 1,8-Dichloroanthracenes." *European Journal of Inorganic Chemistry*, 941–947.

[69] Li, Y., Köse, M. E. & Schanze, K. S. (2013). "Intramolecular Triplet Energy Transfer in Anthracene-Based Platinum Acetylide Oligomers." *Journal of Physical Chemistry B*, *117*, 9025–9033.

[70] Carano, M., Careri, M., Cicogna, F., D'Ambra, I., Houben, J. L., Ingrosso, G., Marcaccio, M., Paolucci, F., Pinzino, C. & Roffia, S. (2001). Electronic Communication in Homobimetallic Anthracene-Bridged η5-Cyclopentadienyl Derivatives of Rhodium(I): Generation and Characterization of the Average-Valence Species $[L_2Rh\{C_5H_4CH_2(9,10\text{-anthrylene})CH_2C_5H_4\}RhL_2]^+$." *Organometallics*, 20, 3478–3490.

In: Advances in Chemistry Research
Editor: James C. Taylor
ISBN: 978-1-53615-828-1

Chapter 4

EFFECT OF CENTRIC GERMANIUM PERTURBATION ON ANTHRACENE

Lemi Türker*

Middle East Technical University, Department of Chemistry,
Üniversiteler, Çankaya/Ankara, Turkey

ABSTRACT

The germanium atom is substituted at three different positions (1, 2 or 9) of anthracene structure. The effect of centric perturbation arising from the replacement of carbon atom by a germanium atom at those positions of anthracene structure has been investigated within constrains of the density functional theory at the levels of B3LYP/6-31++G(d,p) and ωB97X-D/6-31++G(d,p). Various quantum chemical properties, dipole moments, bond lengths, charges, total electronic energies, the highest occupied and lowest unoccupied molecular orbital energies (HOMO and LUMO, respectively) and the interfrontier molecular orbital energy gap values, etc., have been obtained and compared. The germanium atom substitution perturbs the structure but planarity of the system is generally preserved. The isomeric germaanthracenes presently considered are structurally stable having reasonable bond lengths. The calculated infrared and ultraviolet-visible

* Corresponding Author's E-mail: lturker@metu.edu.tr, lturker@gmail.com.

spectra of the perturbed anthracenes are presented. The NICS (0) values of the individual rings in each case have been calculated (B3LYP/6-31++G(d,p) level) and all these results are discussed.

Keywords: anthracene, germaanthracenes, energies, density functional theory, NICS (0)

1. INTRODUCTION

The last quarter of the 20th century evidenced the preparation of π-systems containing novel (p-p) π bonds between carbon and elements from the third, fourth, and fifth main groups of the periodic table. Most of these compounds are predicted to be incapable of existence according to the classical concepts of the double bond. However, (p-p) π bonds between carbon and phosphorus, arsenic, and antimony have been successfully stabilized in resonance-stabilized colored compounds and also in aromatic systems deriving from benzene, naphthalene, and anthracene. Barton and coworkers managed to detect an intermediacy of a compound for an evidence of a germanium-carbon (p-p) π double bond [1]. Germanium-nitrogen double-bond was also reported [2]. Also germanium-carbon (p-p) π bonds have been detected in extremely reactive derivatives of germaethylene [3].

The most common oxidation states for germanium are +4 and +2. The first organogermanium compound prepared was tetraethyl germane ($GeEt_4$)synthesized by the element's discoverer in 1887 [4, 5]. However, the organometallic chemistry of germanium has not been as extensively investigated relative to that of organosilicon or organotin compounds. Germanium prefers to be in the +4 oxidation state, and germylenes, which contain germanium in the +2 oxidation state, often require the use of sterically encumbering or electron donating ligands to enable their isolation and characterization [6].

Early investigations into the synthesis of germanium heterocycles indicated that the chemistry was different from that of silicon.

Several compounds of the type Ge=X (X = C, Ge, N, P, S) have been isolated as monomers. The stabilization of such derivatives, which are generally highly polymerizable, was achieved by using very bulky groups both on the germanium atom and on the heteroelement X. Conjugation (particularly in a few germenes and germaimines) and intramolecular or intermolecular coordination with oxygen or nitrogen, also contributes, in some cases, to the stabilization. [7]. These doubly bonded germanium derivatives are usually thermally stable but must be handled in an inert atmosphere because of their high sensitivity to oxygen and moisture because they are extremely reactive, much more than the corresponding carbon analogues [7-10]. Nakata and coworkers have published various articles about germaaromatic compounds [11-16].

Synthesis of some functional germanium heterocycles were synthesized [17].Digermabenzene complexes appeared to be intermediates to efficient syntheses of digermacyclohexadienes (germirenes) [18-19].

A ladder-type germanium-bridged arene and its donor–acceptor conjugated copolymers were synthesized and its photovoltaic applications investigated [20]. Some unusual germanium(II) calix[4]arene and calix[8]arene complexes were synthesized and their properties etc., were reported [21]. The synthesis of calix[4]arenes with germanium-containing side chains were reported [22]. Li et al., studied weak arene stabilization by studying bulky amido-germanium(II) cation [23].

Some germaanthracene derivatives have found some application as electroluminescence display devices [24]. The synthesis, structure and phosphorescence properties of arene-inserted extended germa[*n*] pericyclynes have been reported [25].

Wang and Wang considered the dimerization reactions of 1-germaanthracene, 2-germaanthracene and 9-germaphenanthrene theoretically [26]. Sigolaev et al., performed some quantum chemical calculations on germaanthracenes [27].

2. METHOD OF CALCULATION

Geometry optimizations of all the structures leading to energy minima were initially achieved by using MM2 method, followed by semi-empirical PM3 self-consistent fields molecular orbital (SCF MO) method [28, 29] at the restricted level [30, 31]. Subsequent optimizations were achieved at Hartree-Fock level by using various basis sets hierarchically. Then, geometry optimizations were managed within the framework of density functional theory [32, 33], finally at the level of RB3LYP /6-31++G(d,p) [30] and ωB97X-D/6-31++G(d,p) [34, 35]. The exchange term of B3LYP consists of hybrid Hartree–Fock and local spin density (LSD) exchange functions with Becke's gradient correlation to LSD exchange [33, 36]. Note that the correlation term of B3LYP consists of the Vosko, Wilk, Nusair (VWN3) local correlation functional [37] and Lee, Yang, Parr (LYP) correlation correction functional [38]. Note that ωB97X-D functional provides excellent geometries and energies. It has high HF percentage (the second as a long-range corrected functional), which helps mitigate the self-interaction error and the difficult excitation energies required for charge transfer reactions [34, 35].

The vibrational analyses were also done. The total electronic energies are corrected for the zero point vibrational energy (ZPE). The stationary points to energy minima were proved in all the cases by calculation of the second derivatives of energy with respect to the atom coordinates. The normal mode analysis for each structure yielded no imaginary frequencies for the $3N-6$ vibrational degrees of freedom, where N is the number of atoms in the system. This indicates that the structure of each molecule corresponds to at least a local minimum on the potential energy surface. All these calculations were done by using the Spartan 06 package program [39].

For NICS data, the absolute NMR shielding values [40] were calculated by employing the Gauge-Independent Atomic Orbital method [41] with the restricted closed shell formalism at the level of B3LYP/6-31++G(d,p). The NICS values were collected by calculating absolute NMR shielding values at the ring centers, NICS (0). The NICS (0) calculations of the present systems were performed by the use of the Gaussian 03 package program [42].

3. Results and Discussion

Various carbon group elements (IVA) having multiple p_π-p_π bond are known, including C=Ge linkage [43]. In these structures, the following resonance forms are assumed where Me stands for a heteroatom. Delocalization of electron density of a carbanion center over the system of conjugated bonds is one of the possible ways to stabilize such structures [43]. For example, 6,6-dimethyl-6-germafulvene was shown to have enhanced stability by quantum chemical calculations [38].

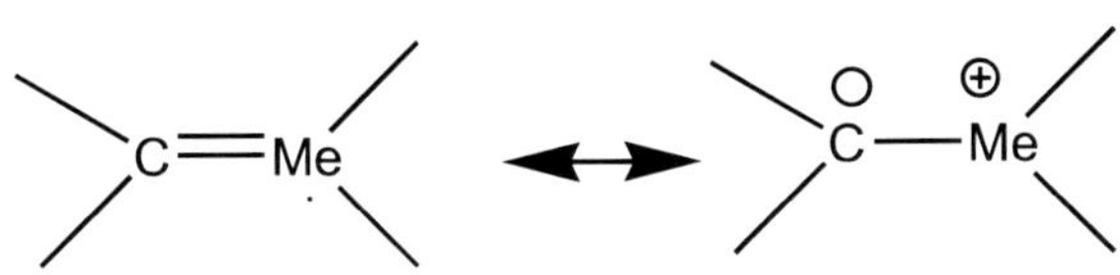

3.1. Structures

Figure 1 shows the optimized structures of germaanthracenes ($C_{13}H_{10}Ge$) considered presently. They have Ge atom at centers 1, 2 or 9 positions of anthracene structure. The figure also shows the direction of the dipole moment vectors. The perturbation caused by the germanium substitution in anthracene, structurally does not disturb the planarity. Table 1 shows some calculated properties of them. Note that the ovality of each structure is the same in each level of calculations. Each area and volume values at the ωB97X-D/ 6-31++G(d,p) level is either equal or less than the corresponding value of B3LYP/6-31++G(d,p) level.

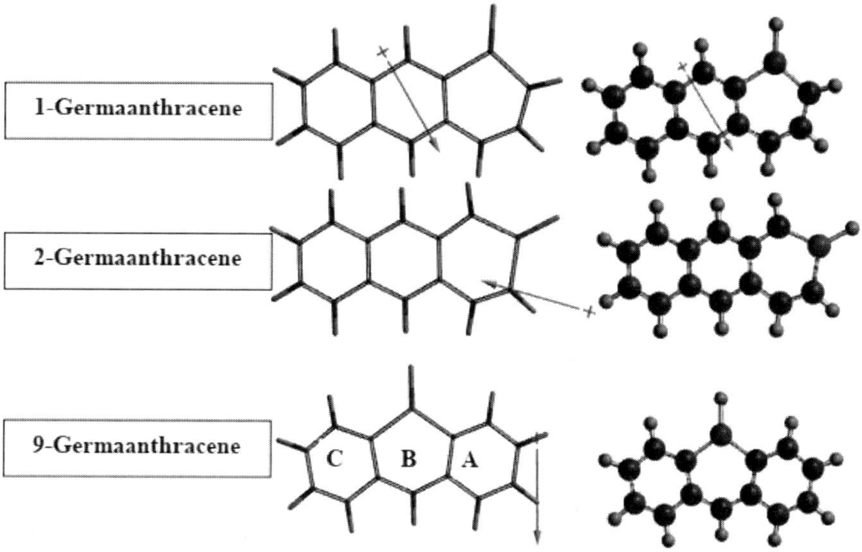

Figure 1. Optimized structures of the anthracenes considered (B3LYP /6-31++G(d,p)).

Table 1. Some properties of the germaanthracenes considered

Structure	Area (Å^2)	Volume (Å^3)	Ovality	Polarizability
1-Germaantracene	220.73 (219.90)	212.27 (211.47)	1.28 (1.28)	57.88 -
2-Germaanthracene	220.08 (220.08)	212.07 (212.07)	1.28 (1.28)	57.82 -
9- Germaanthracene	220.92 (220.05)	212.37 (211.55)	1.28 (1.28)	57.95 -
B3LYP/6-31++G(d,p) level.The values in parenthesis are ωB97X-D/ 6-31++G(d,p) level output. All belong to C_1 point group.				

Table 2 displays the dipole moment components and dipole moments for anthracene and germaanthracenes considered. Compared to anthracene case the X-component in 1-germaantracene; Y-and Z-components in 2-germaantracene and 9-germaantracene are highly perturbed. The magnitude order of the resultant dipole moments is 2-germaanthracene>1-germaanthracene>9-germaanthracene>anthracene.

Table 2. Dipole moment components and dipole moments for anthracene and germaanthracenes considered

Structure	X	Y	Z	Dipole moment
Anthracene	0.000164	-0.001278	0.000996	0.001629
1-Germaantracene	-0.409664	-0.001578	0.704935	0.815328
2- Germaanthracene	-0.000405	-0.213850	-0.870132	0.896026
9- Germaanthracene	0.000056	-0.594912	0.334200	0.682357
Debye unit. B3LYP/6-31++G(d,p) level.				

Some thermodynamic properties of the germaanthracenes considered are displayed in Table 3. The heat of formation values (ΔH_f°) calculated at the level of PM3// B3LYP/6-31++G(d,p) indicate that all these structure are endothermic. The heat of formation values follow the order of 2-

Table 3. Some thermodynamic properties of the germaanthracenes considered

Structure	ΔHf°	H°	G°	S°	Cv
1-Germaantracene	397.398 (390.813)	-6768561.284	-6768685.104	415.27	153.97
2-Germaanthracene	406.968 (406.968)	-6768558.764	-6768682.19	413.91	154.62
9-Germaanthracene	381.971 (374.451)	-6768558.396	-6768682.321	415.69	152.33
B3LYP/6-31++G(d,p) level. H°, G° and ΔHf° values are in kJ/mol. S° and Cv values in J/mol°. ΔHf° values at the level of PM3// B3LYP/6-31++G(d,p). The values in parenthesis are at the level of PM3// ωB97X-D/ 6-31++G(d,p).					

Table 4. Various energies of germaanthracenes (B3LYP/6-31++G(d,p))

Structure	E	ZPE	Ec
1-Germaantracene	-6769008.81	480.80	-6768528.01
2- Germaanthracene	-6769006.13	480.91	-6768525.22
9- Germaanthracene	-6769007.02	481.88	-6768525.14
Energies in kJ/mol.			

Germaanthracene >1-Germaantracene > 9-Germaanthracene. Thus, according to the results the first and third ones are the most and least endothermic structures, respectively.

Table 5. Various energies of germaanthracenes (ωB97X-D/ 6-31++G(d,p))

Structure	E	ZPE	Ec
1-Germaantracene	-6768641.098	488.24	-6768152.858
2- Germaanthracene	-6768638.115	488.45	-6768149.665
9- Germaanthracene	-6768638.275	488.96	-6768149.315
Energies in kJ/mol.			

Table 4 shows the total electronic energy (E), zero point vibrational energy (ZPE) and the corrected total electronic energy (E_c). The order of E_c values is 1-germaanthracene < 2- germaanthracene < 9-germaanthracene. So, 1-germaanthracene should be more stable than the others at B3LYP/6-31++G(d,p) level of calculations. The same order of stability is expected as the results of ωB97X-D/ 6-31++G(d,p)level of calculations (Table 5). Note that in each case, the basis set used is the same but the functional differs.

3.2. Bond Lengths and Bond Angles

Figure 2 shows the bond lengths calculated at B3LYP/6-31++G(d,p) level. The carbon-germanium bond lengths vary at each side of the germanium. The deviation is greater in 1-germaantracene and minimum in 9-germaanthracene. Germanium-carbon single bond length is 1.945 Å in H_3GeCH_3 and 1.896 Å in H_3GeCCH [44] and generally in the range of 1.90-2.05 Å. Whereas, a typical Ge=C bond length falls into 1.80-1.83 Å [45]. In the present structures germanium-carbon bond lengths vary about 1.81-1.88 Å. So, some Ge-C bonds are in between double and single bond character.

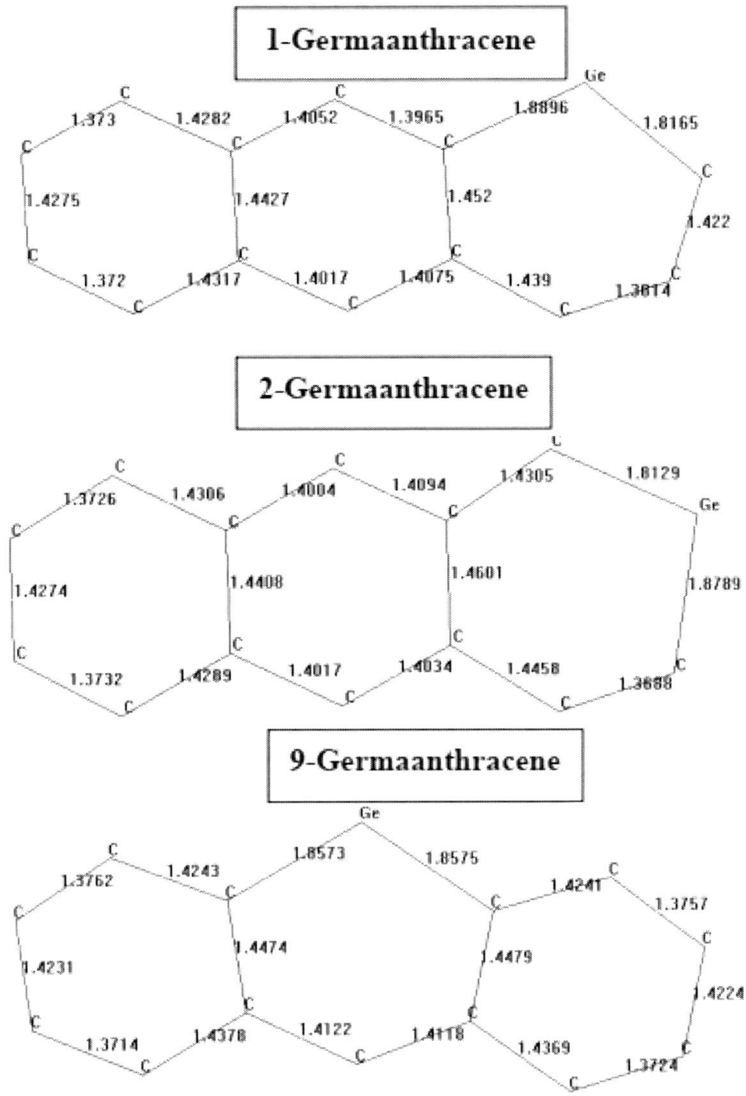

Figure 2. Bond lengths (Å) of germaanthracenes considered (B3LYP /6-31++G(d,p), hydrogens not shown).

The C-Ge-C bond angles are 108.28°, 107.32° and 109.20° in 1-germaantracene, 2- germaanthracene and 9-germaantracene, respectively. Since C-Ge bond lengths are longer than the C-C bond lengths, consequently, the linearity of the rings compared to anthracene has been somewhat disturbed in 9-germaantracene case. The carbon-carbon bond lengths show some bond length alternation implying some annulenic behavior of the rings (conversely less aromatic!) [46].

3.3. Charges

Figure 3 displays the electrostatic charges (eu) on atoms of the structures considered. They are ESP charges. The ESP charges are obtained by the program based on a numerical method (CHELP algorithm) that generates charges that reproduce the electrostatic potential field from the entire wavefunction [39]. In this algorithm the charges at the atoms are chosen to best describe the external field surrounding the molecule. Ideally this area should include everything outside of the Van der Waal radii. In practice, a shell surrounding the atoms having a thickness of 5.5 au is chosen for ESP calculations.

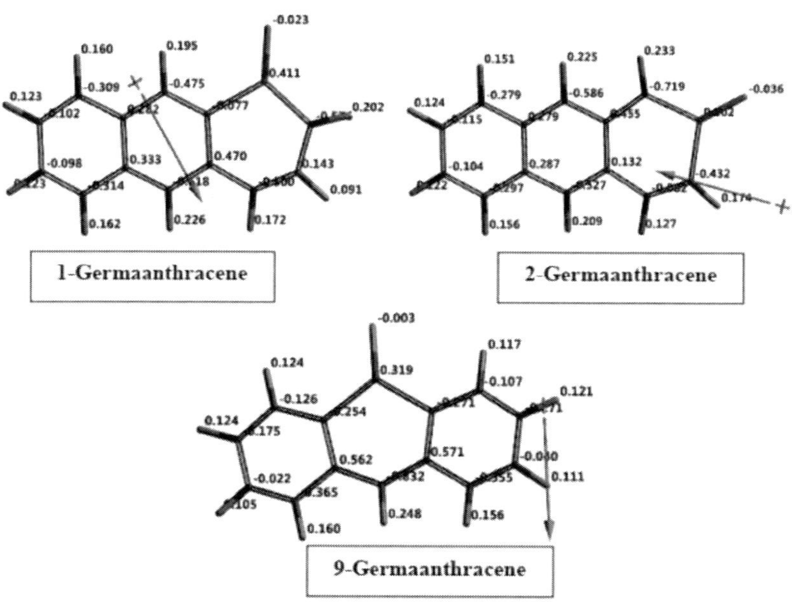

Figure 3. Electrostatic charges (ESP) on the germaanthracenes considered (B3LYP /6-31++G(d,p)).

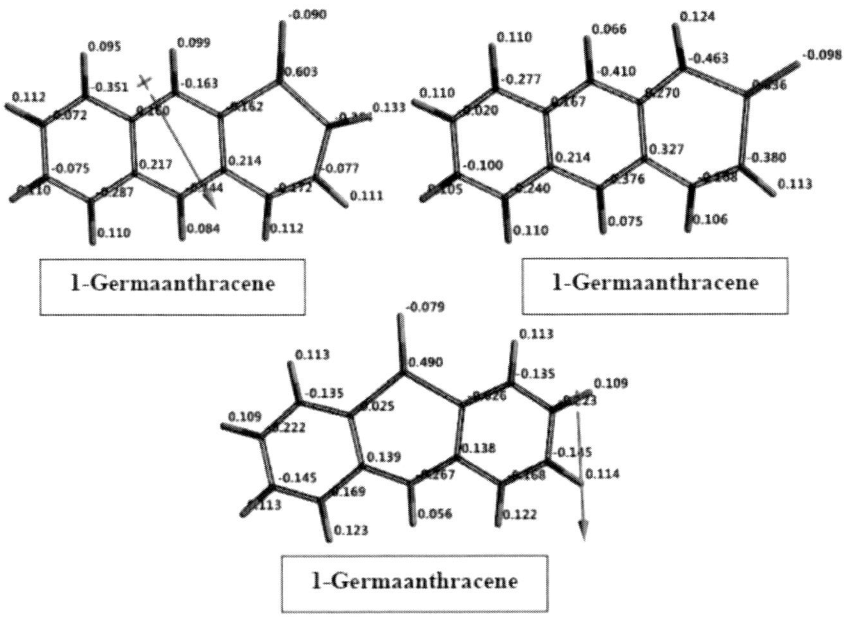

Figure 4. Mulliken charges on the germaanthracenes considered (B3LYP /6-31++G(d,p)).

In each case the Ge atom acquires some positive partial charge having the order of magnitudes as 2-germaanthracene > 1-germaanthracene > 9-germaanthracene. Atoms around the germanium (C and H atoms) are all partially negatively charged. Also all the remaining hydrogens are positively charged in all the cases. The carbon atoms are mostly negatively charged.

The calculated Mulliken charges on the structures (displayed in Figure 4) also indicate some positive charge development on the Ge atoms and some negative charges on the neighbor atoms.

3.4. IR Spectra

Figure 5 shows the calculated IR spectra of anthracene and germaanthracenes considered. All the germaanthracenes have a characteristic peak in the region of 2135-2147 cm^{-1} which stands for Ge-H stretching.

1-germaanthracene and 9-germaanthracene have an additional peak in between 1500-1250 cm^{-1} which is C-H bending. Due to the asymmetry in 1- and 2-germaanthracenes, the symmetric and asymmetric C-H stretchings which occur in the region of 3000-3300 cm^{-1} cover larger part of the spectrum as compared to 9-germaanthracene case.

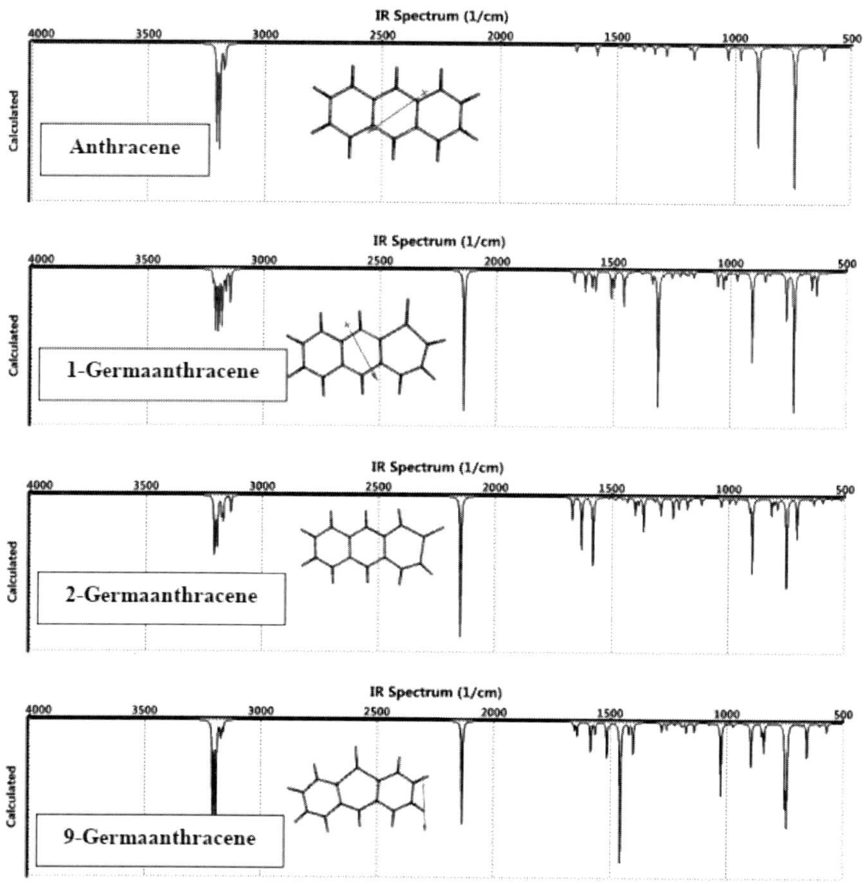

Figure 5. IR spectra of the germaanthraces considered (B3LYP /6-31++G(d,p)).

3.5. Molecular Orbital Energies and UV Spectra

Figure 6 is the molecular orbital energy spectra for the structures considered. The HOMO energy level of all the germaanthracenes is higher than the HOMO level of anthracene having the order of anthracene < 2-germaanthracene < 1-germaanthracene < 9-germaanthracene. Whereas, all the LUMO energy levels are lower than the respective level of anthracene following the sequence of 9-germaanthracene < 1-germaanthracene < 2-germaanthracene < anthracene (Table 6).

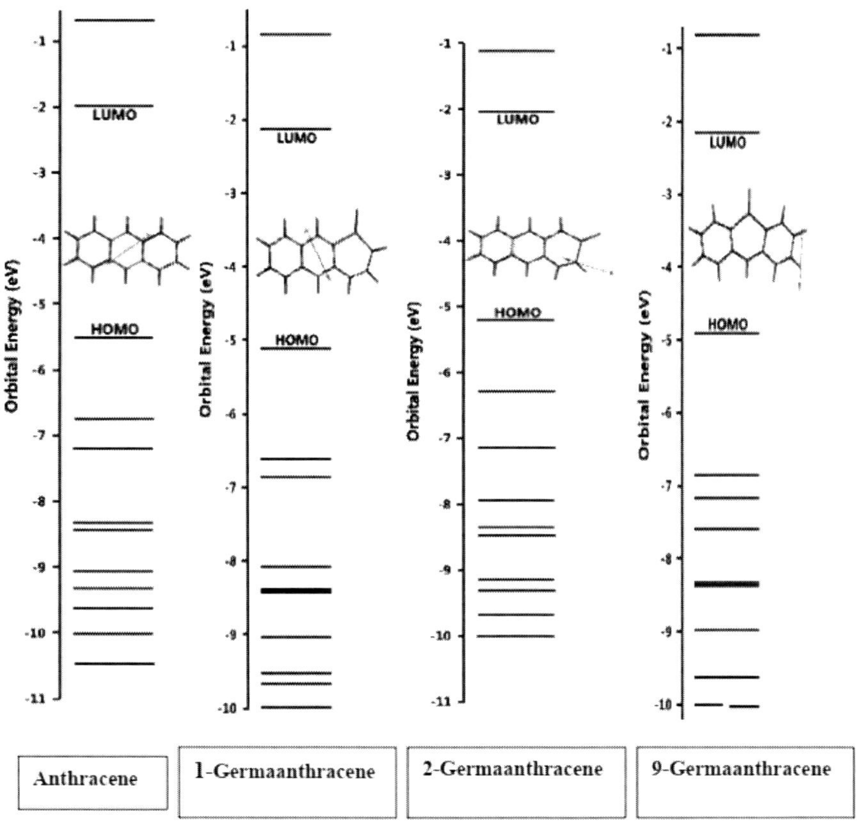

Figure 6. Some of the molecular orbital energy levels of the anthracenes considered (B3LYP /6-31++G(d,p)).

Table 6. The HOMO, LUMO energies and interfrontier molecular orbital energy gaps (Δε)

Structure	HOMO	LUMO	Δε
Anthracene	-532.27	-190.76	341.51
1-Germaantracene	-493.98	-204.92	289.06
2- Germaanthracene	-503.00	-195.42	307.58
9- Germaanthracene	-474.66	-207.59	267.07
Energies in kJ/mol. B3LYP/6-31++G(d,p) level.			

Perturbation molecular orbital theory (PMO) [47] estimates the order of perturbed molecular orbital energies simply from the molecular orbital energies and molecular orbital coefficients of the parent system, (anthracene) within the framework of the Hückel molecular orbital theory (HMO) by using the formula;

$$\delta\varepsilon_j = c_{j\rho}{}^2\delta\alpha \qquad (1)$$

where $\delta\varepsilon_j$ is the energy of the jth molecular orbital of the perturbed system, $c_{j\rho}$ is the molecular orbital coefficient of the perturbed atom ρ in the jth molecular orbital and δα is the Coulomb integral change by replacing carbon atom with a heteroatom (eg; Ge). It is related to electro negativity difference of the atoms considered in the centric perturbation. The HOMO coefficient of anthracene at centers 1, 2 and 9 are 0.311, 0.220 and -0.440, [48] respectively within HMO theory. If germanium atom exerts electron withdrawing effect, both the HOMO and LUMO levels are expected to be lowered as compared to anthracene. If it donates electrons the respective levels should rise. On the other, hand, in even alternant systems (e.g.,; anthracene) as the extended conjugation happens, the HOMO-LUMO energy gap decreases [49]. So, the frontier molecular orbital energy gap (ε_{LUMO}-ε_{HOMO}) order of 9- germaanthracene < 1-germaanthracene < 2-germaanthracene that is 9-germaanthracene has the narrowest gap should have better extended conjugation. Indeed, UV-VIS spectrum of 9-germaanthracene shows some bathochromic effect compared to anthracene and 2-germaanthracene (see Figure 7). Note that spectrums of 1-

germaanthracene and 9-germaanthracene look similar at the first sight, but 9-germaanthracene exhibits bathochromic effect at all the λ_{max} values.

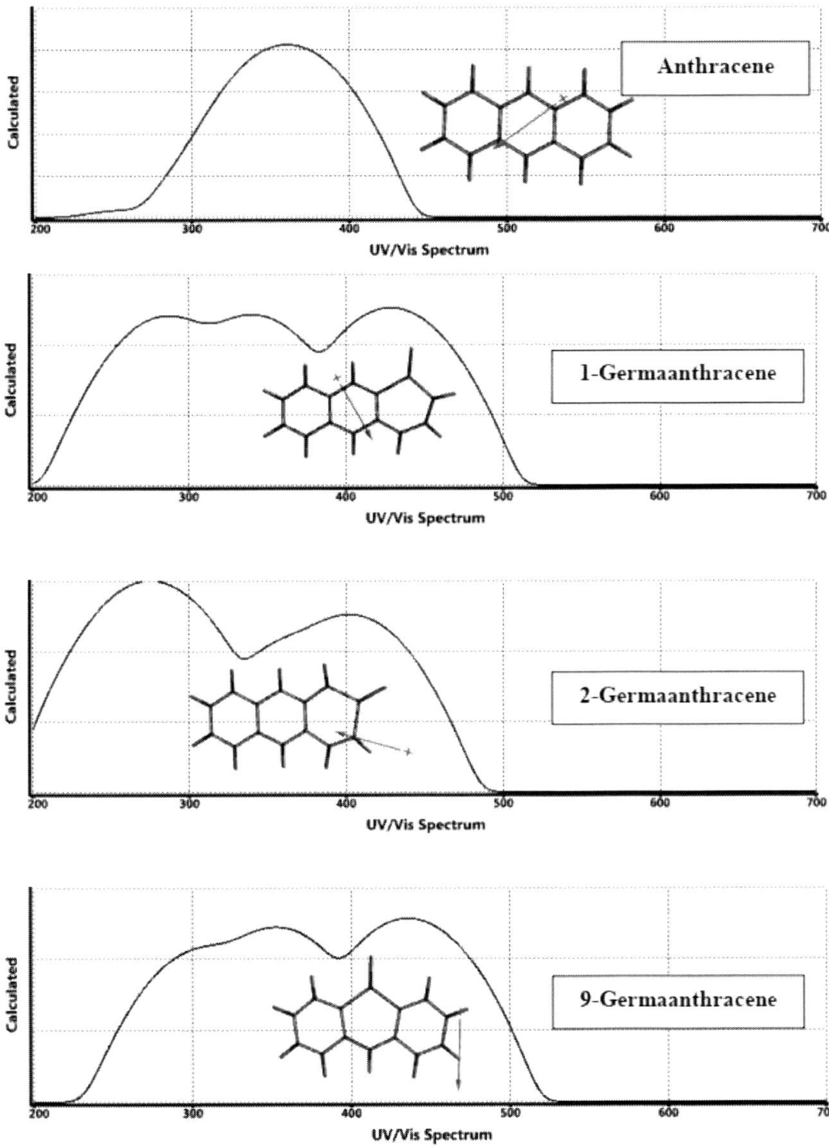

Figure 7. UV-VIS spectra of the anthracenes considered (B3LYP /6-31++G(d,p)).

Table 7. The λ_{max}values and their relative intensities

Structure	λmax	Relative intensity
Anthracene	360.86	1.00
1-Germaantracene	286.62 342.03 428.81	0.50 0.50 1
2- Germaanthracene	275.13 366.47 403.00	1 0.02 0.11
9- Germaanthracene	305.50 353.60 436.17	0.08 0.56 1
B3LYP/6-31++G(d,p) level.		

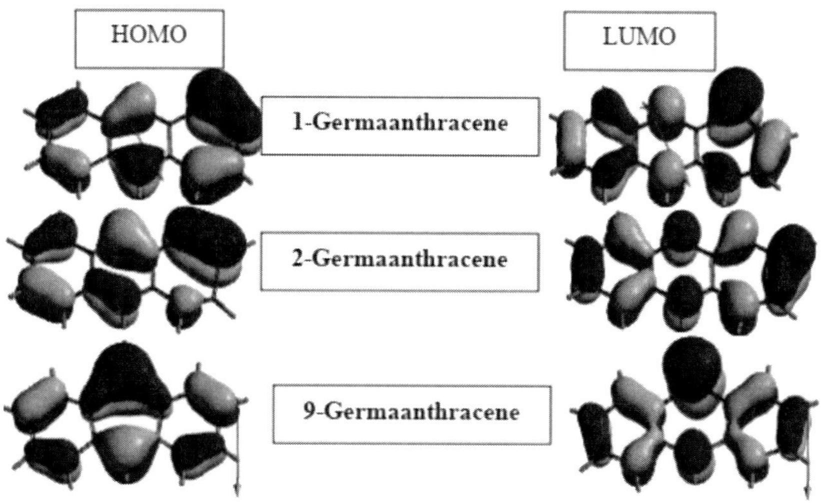

Figure 8. The HOMO and LUMO pattern of the germaanthracenes considered (B3LYP /6-31++G(d,p)).

Figure 7 shows the UV-VIS spectra of the structures (the time-dependent DFT). Table 7 displays λ_{max} values and their relative intensities of concern. As seen from Figure 7 substitution of germanium atom perturbs the anthracene structure so that spectrum expands towards the ultraviolet region in all the germaanthracenes but in different extents.

Figure 8 displays the HOMO and LUMO patterns of the structures. As seen from the figure germanium atom largely contributes into the HOMO and LUMO. Some of the carbon atoms which reside on the fusion points of the rings contribute nil into the HOMO and LUMO (frontier molecular orbitals).

3.6. Electrostatic Potential and LUMO Maps

The electrostatic potential maps of the structures considered are shown in Figure 9. As seen there, germanium site of the structures coincides with relatively more positive electrostatic potential compared to periphery of the structures which contains negatively charged carbon atoms.

Figure 10 shows the LUMO maps of the structures obtained at the level of B3LYP /6-31++G(d,p). The LUMO maps encodes the (absolute value) of the LUMO on to an electron density surface. This reveals where an electron pair (a nucleophile) is likely to attack [50].

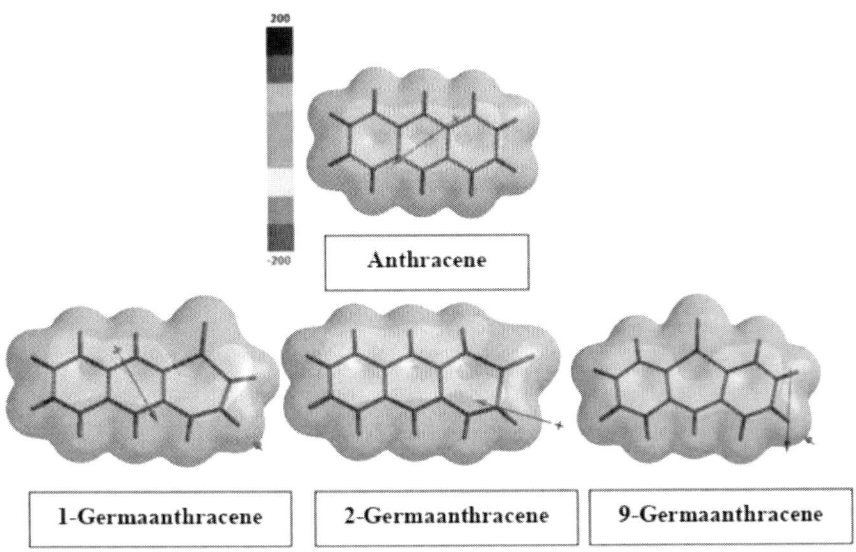

Figure 9. The electrostatic potential maps of the anthracenes considered (B3LYP /6-31++G(d,p)).

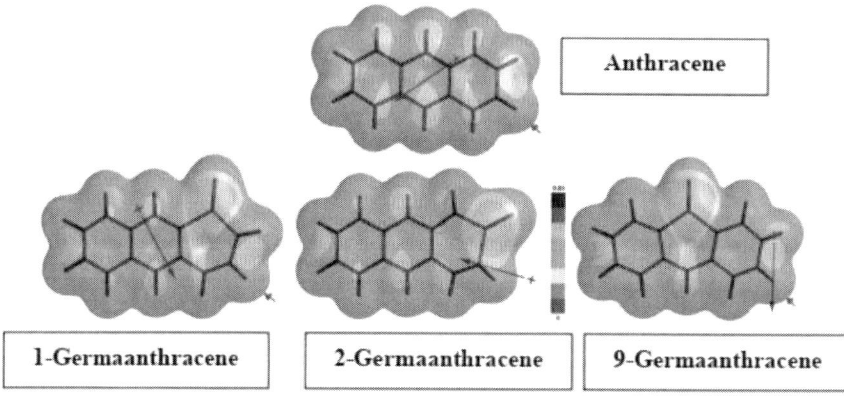

Figure 10. LUMO maps of the structures considered (B3LYP /6-31++G(d,p)).

3.7. NICS

Aromaticity arises by a combination of various properties in cyclic delocalized systems. In general, aromaticity is discussed in terms of energetic, structural and magnetic criteria [51-56]. Schleyer, in 1996, has introduced a simple and efficient probe for aromaticity that is "nucleus-independent chemical shift" (NICS) [57]. It is the computed value of the negative magnetic shielding at some selected point in space which is generally, at a ring or cage center. The calculated data piled in the literature indicate that negative NICS values denote aromaticity (-11.5 for benzene, -11.4 for naphthalene) contrary to positive NICS values which denote antiaromaticity (28.8 for cyclobutadiene) while small NICS values indicate non-aromaticity (-2.1 for cyclohexane, -1.1 for adamantane). NICS may be a useful indicator of aromaticity that usually correlates successfully with the other energetic, structural and magnetic criteria for aromaticity [58-61]. Resonance energies and magnetic susceptibilities are measures of the overall aromaticity of a polycycle, but do not provide information about the individual rings. In contrast, NICS has been proved to be an effective probe for local aromaticity of individual rings of polycyclic systems.

Table 8. NICS (0) values of the rings in anthracene and germaanthracenes considered

Structures	Ring-A	Ring-B	Ring-C
Anthracene	-7.3539	-10.8975	-7.3539
1-Germaantracene	-6.5729	-10.7137	-7.4811
2-Germaantracene	-6.3450	-10.4760	-7.3004
9-Germaantracene	-7.1029	-9.5109	-6.9893
B3LYP/6-31++G(d,p) level of calculations. The rings are named from left to right (see Figure 1).			

Table 8 shows NICS (0) values of the rings in the presently considered structures. In each case the results show more aromatic character for ring-B than the others. The presently obtained NICS(0) value, indicating odd behavior of anthracene, is not due to the failure of B3LYP/6-31++G(d,p) basis set only, but even B3LYP/6-31G(d) level of calculations yield the NICS(0) values of anthracene as -7.3539, -10.9047 and -7.3539, respectively for rings A-C. All these results are controversial to some other aromaticity tests applied on anthracene structure [57, 62-64]. However, when the analysis of these structures is carried out ring-wise it might give some insight. Then, the order of aromaticity of A-type rings is anthracene > 9-germaanthracene >1-germaanthracene > 2-germaanthracene; for B-type rings anthracene > 1-germaanthracene > 2- germaanthracene > 9- germaanthracene; for C-type 1-germaanthracene > anthracene > 2-germaanthracene >9-germaanthracene. Aromaticity of A-type ring in 9- germaanthracene is comparable but less than the same type of ring in anthracene. This could be justified because both systems have the same number of Clar's sextets [65] and have A-type rings consisting of only carbon atoms which are not the case in 1-germaanthracene and 2- germaanthracene. Note that A- and C-type rings in 9-germaanthracene are not equivalent in contrast to anthracene which could be the cause of C1 symmetry possessed in the perturbed structure.

CONCLUSION

The present treatise on anthracene as a result of monocentric germanium perturbation have revealed that the resultant systems are stable with reasonable C-Ge(p-p)π bonds, where the germanium atom acquires some partial positive charge (the ESP and Mulliken charges). The structures are planar, having an imperfect germanium containing hexagonal ring. The stability order of these isomeric structures indicates that the most and least stable ones are 1- and 9-germaantracenes, respectively. The perturbation effect of germanium also affects the molecular orbital energy levels and the FMO gaps follow the order of 9- < 1- < 2-germaantracene.

On the other hand, the NICS calculations of these germaanthracenes yield the middle ring to be more aromatic than the others which is somewhat doubtful. However, this is not specific to germaanthracenes considered, but also the same type result was obtained for anthracene itself.

REFERENCES

[1] Barton, T. J., Kline, E. A. and Garvey, P. M. (1973). Evidence for a germanium-carbon(p-p) pi double bond. *J Am Chem. Soc.95*(9): 3078–3078.

[2] Glidewell, C. (1990). Germanium-carbon versus germanium-nitrogen double-bond formation in the reactions between germylenes and diazo compounds. *Structural Chemistry*. 1(4): 317–323.

[3] Jutzi, P. (1975). 9-Arsaanthracene. *Angewandte Chemie*. 14(4):232-245.

[4] Winkler, C. (1886). Germanium, Ge, ein neues, nichtmetallisches [Germanium, Ge, a new, non-metallic]. *Element. Chem. Ber.* (19):210-211.

[5] Winkler, C. J. (1887). Mittheilungen über das Germanium [Communications about Germanium]. *Prakt.. Chem. 36* (1): 177-209.

[6] Durant, P. J. and Durant, B. (1970). *Introduction to advanced inorganic chemistry.* London, Longman.

[7] Escudié, J., Couret, C., Ranaivonjatovo, H. and Satge, J. (1994). Recent developments in the chemistry of stable doubly bonded germanium compounds. *Coordination Chemistry Reviews.* 130 (1–2): 427-480.

[8] Bickelhaupt, F., Jongsma, C., Koe, D.P., Lourens, R., Mast, N.R., Van Mourik, G. L., Vermeer, H. and Weustink, R. J. M. (1976). Synthesis and application of precursors of hetero-anthracenes: 2-halo- and 2,2'-dihalodiphenylmethanes and methine substituted derivatives. *Tetrahedron.* 32(15): 1921-30.

[9] Sasamori, T., Inamura, K., Hoshino, W., Nakata, N., Mizuhata, Y., Watanabe, Y., Furukawa, Y. andTokitoh, N. (2006). Synthesis and characterization of two isomers of14π-electron germaaromatics: kinetically stabilized 9-germaanthracene and 9-germaphenanthrene. *Organometallics.* 25(15): 3533-3536.

[10] Mizuhata, Y., Fujimori, S., Sasamori, T. and Tokitoh, N. (2017). Germabenzenyl potassium: A germanium analogue of a phenyl anion. *Angew Chem Int Ed Engl.* 56(16): 4588-4592.

[11] Nakata, N., Takeda, N., Tokitoh, N. (2002). Synthesis and properties of thefirst stable germabenzene. *J Am Chem. Soc.* 124 (24): 6914-6920.

[12] Nakata, N., Takeda, N., Tokitoh, N. (2003). Reaction of a stable germabenzene withchalcogens: Synthesis and structure of a novel germanium analog of pentathiepane, 1,2,3,4,5,6-pentathiagermepane. *Journal of Organometallic Chemistry.* 672 (1-2):66-71.

[13] Nakata, N., Takeda, N., Tokitoh, N. (2003). η^6-Germabenzene complexes of chromium and molybdenum. *Angewandte Chemie – International Edition*, 42 (1): 115-117.

[14] Nakata, N., Takeda, N., Tokitoh, N. (2001). Synthesis and structure ofkineticallystabilized 2-germanaphthalene: The first stable neutral germaaromatic compound. *Organometallics*, 20 (26): 5507-5509.

[15] Nakata, N., Takeda, N., Tokitoh, N. (2002). Reactions of 2-germanaphthalene with elemental sulfur and selenium: Synthesis of

novel cyclic polychalcogenides containing a germanium, trichalcogenagermolanes. *Chemistry Letters*. (8): 818-819.

[16] Nakata, N., Takeda, N., Tokitoh, N. (2003). Synthesis and properties of the firststable neutral germaaromatic compound, 2-{2,4,6-Tris-[bis(trimethylsilyl)methyl]phenyl)-2-germanaphthalene. *Organometallics*. 22 (3):481-489.

[17] Scibelli, J.V. and Curtis, M.D. (1972). Synthesis of some functional germaniumheterocycles: Germirenes and derivatives of germacyclopentene. *Journal of Organometallic Chemistry*. 40: 317-325.

[18] Volpin, M. E., Dulova, V.G., Struchkov, Y.T., Bokiy, N. K. and Kursanov, D. N.J. (1967). Structure and properties of unsaturated germanium heterocycles. *Orgonometal. Chem.* 8: 87-96.

[19] Johnson, F. and Gohlke, R.S. (1962). On the nature of "germirene" derivatives. *Tetrahedron Lett.* 26: 1291- 1295.

[20] Jwo, P.C., Lai, Y.Y., Tsai, C.E., Lai, Y.Y., Liang, W.W., Hsu, C.S. and Cheng, Y.J. (2014). A new ladder-type germanium-bridged dithienocarbazole arene and its donor–acceptor conjugated copolymers: Synthesis, molecular properties, and photovoltaic applications. *Macromolecules*. 47(21): 7386–7396.

[21] Wetherby, A.E., Goeller, L.R., Pasquale, D.A.G., Rheingold, A.L. and Weinert, C. S. (2007). Synthesis and structures of an unusual germanium (II) calix[4]arene complex and the first germanium(II) calix[8]arene complex and their reactivity with diironnonacarbonyl. *Inorganic Chemistry*. 46(18): 7579-86.

[22] Sakurai, T. and Takeuchi, Y. Materials, (2004). Synthesis of calix[4]arenes modified with germanium - containing side chains on the upper rim. *Applied Organometallic Chemistry*. 18(1): 23-27.

[23] Li, J., Schenk, C., Winter, F., Scherer, H., Trapp, N., Higelin, A., Keller, S., Pöttgen, R., Krossing, I. And Jones, C. (2012). Weak arene stabilization of bulkyamido - germanium(II) and Tin(II) Monocations. *Angewandte Chemie*. 51(38):9557-9561.

[24] Park, J.W. and Lee, J.H. (2006). *Compounds and organic electroluminescence display device comprising the same*, PCT Int. Appl. WO 2006065105 A1 20060622.
[25] Tanimoto, H., Mori, J., Ito, S., Nishiyama,Y., Morimoto, T., Tanaka, K., Chujo, Y. and Kakiuchi, K. (2017). Arene - ınserted extended germa[*n*]pericyclynes: synthesis, structure, and phosphorescence properties. *Chemistry A new European Journal*. 23(42):10080-10086.
[26] Wang, Y., and Wang, Y.P. (2013). Theoretical study on dimerization reactions of 1-germaanthracene, 2-germaanthracene and 9-germaphenanthrene. *Huaxue Yanjiu Yu Yingyong*. 25(2): 189-193.
[27] Sigolaev, Y.F., Semenov, S.G. and Reikhsfel'd, V.O. (1983). Quantum-chemical investigation of the interaction of heteroatoms and phenyl groups in the molecules of 10,10-dichloro-10-germa-9-oxa-9, 10-dihydroanthracene and 10,10-dimethyl-10-germa-9-thia-9,10-dihydroanthracene. *Theoretical and Experimental Chemistry*. 19(3): 322–324.
[28] Stewart, J.J.P. (1989). Optimization of parameters for semiempirical methods I. Method. *J Comput Chem*. 10: 209-220.
[29] Stewart, J.J.P. (1989). Optimization of parameters for semiempirical methods II. Applications. *J Comput Chem*.10: 221-264.
[30] Leach, A.R. (1997). *Molecular modeling*. Essex, Longman.
[31] Fletcher, P. (1990). *Practical methods of optimization*. New York, Wiley.
[32] Kohn, W. and Sham, L. (1965). Self-consistent equations including exchange and correlation effects. *J Phys Rev*. 140:1133-1138.
[33] Parr, R.G. and Yang, W. (1989). *Density functional theory of atoms and molecules*. London, Oxford University Press.
[34] Minenkov, Y., Singstad, A., Occhipinti, G. and Jensen, V.R. (2012). The accuracy of DFT-optimized geometries of functional transition metal compounds: a validation study of catalysts for olefin metathesis and other reactions in the homogeneous phase. *Dalton Trans*. 41(18): 5526-5541.
[35] Kozuch, S. and Martin. J.M.L. (2013). Halogen bonds: Benchmarks and theoreticalanalysis. *J Chem Theory Comput*. 9: 1918−1931.

[36] Becke, A.D. (1988). Density-functional exchange-energy approximation with correct asymptotic behavior. *Phys Rev A*. 38: 3098-3100.

[37] Vosko, S.H., Vilk, L. and Nusair, M. (1980). Accurate spin-dependent electron liquid correlation energies for local spin density calculations: a critical analysis. *Can J Phys*., 58: 1200-1211.

[38] Lee, C., Yang, W. and Parr, R.G. (1988). Development of the Colle-Salvetti correlation-energy formula into a functional of the electron density. *Phys Rev B*. 37. 785-789.

[39] *SPARTAN 06*. (2006). Wavefunction Inc., Irvine CA, US.

[40] Pulay, P.J., Hinton, F.andWolinski, K. (1993). *Nuclear magnetic shieldings and molecular structure*. Eds.: Tossel, J.A. NATO ASI Series C, vol. 386, Netherlands, Kluwer, pp. 243.

[41] Hehre, W.J., Radom, L., Schleyer, P.R. and Pople, J.A. (1986).*Ab initio molecular orbital theory*. New York, Wiley.

[42] Frisch, M.J., Trucks, G.W., Schlegel, H.B., Scuseria, G.E., Robb, M.A., Cheeseman, J.R., Montgomery, Jr., J.A., Vreven, T., Kudin, K.N., Burant, J.C., Millam, J.M., Iyengar, S.S., Tomasi, J., Barone, V., Mennucci, B., Cossi, M., Scalmani, G., Rega, N., Petersson,G.A., Nakatsuji, H., Hada, M., Ehara, M., Toyota, K., Fukuda, R., Hasegawa, J., Ishida,M., Nakajima, T., Honda, Y., Kitao, O., Nakai, H., Klene, M., Li, X., Knox, J. E., Hratchian, H. P., Cross, J.B., Bakken, V., Adamo, C., Jaramillo, J., Gomperts, R., Stratmann, R.E., Yazyev, O., Austin, A.J., Cammi, R., Pomelli, C., Ochterski, J.W., Ayala, P.Y., Morokuma, K., Voth, G.A., Salvador, P., Dannenberg, J.J., Zakrzewski, V.G., Dapprich, S., Daniels, A.D., Strain, M.C., Farkas, O., Malick, D.K., Rabuck, A.D., Raghavachari, K., Foresman, J.B., Ortiz, J.V., Cui, Q., Baboul, A.G., Clifford, S.,Cioslowski, J., Stefanov, B.B., Liu, G., Liashenko, A., Piskorz, P., Komaromi, I., Martin,R. L., Fox, D.J., Keith, T., Al-Laham, M.A., Peng, C.Y., Nanayakkara, A., Challacombe, M., Gill, P.M.W., Johnson, B., Chen, W., Wong, M.W., Gonzalez, C., and Pople, J.A., Gaussian, Inc., Wallingford CT, (2004.

[43] Beletskaya, I.P. and Zemlyansky, N.N. (1984). Carbon-group IVA element compoundswith multiple p_π-p_π bond. New approaches to

synthesis. In Reutov OA. (Ed.) *Advances inorganometallic chemistry*. Moscow, Mir.

[44] Vilkov, L.V., Mastryukov, V.S. and Sadova, N.I. (1983). *Determination of the geometrical structure of free molecules*. Moscow, Mir.

[45] Baines, K.M. and Stibbs, W. G. (1995). The molecular structure of organogermanium compounds. *Coordination Chemistry Reviews*. 145: 157-200.

[46] Dewar, M.J.S. and Dougherty, R.C. (1975). *PMO theory of organic chemistry*. NY, Plenum-Rosetta.

[47] Dewar, M.J.S. (1969). *The molecular orbital theory of organic chemistry*. New York, McGraw- Hill.

[48] Heilbronner, E. and Bock, H. (1976). *The HMO model and its application*. Vol. 3. NY, Wiley.

[49] Fleming, I. (1976). *Frontier orbitals and organic chemical reactions*. NY, Wiley.

[50] Spartan. (2005). *Molecular modeling in physical chemistry*. Irwine, Wavefunction Inc.

[51] Minkin, V.I., Glukhovtsev, M.N. and Simkin, B.Y. (1994). *Aromaticity and antiaromaticity: Electronic and structural aspects*. New York, Wiley.

[52] Schleyer, P.R. and Jiao, H. (1996). What is aromaticity? *Pure Appl Chem*. 68: 209-218.

[53] Glukhovtsev, M.N. (1997). Aromaticity today: energetic and structural criteria. *J Chem. Educ*. 74: 132-136.

[54] Krygowski, T.M. Cyranski, M.K. Czarnocki, Z., Hafelinger, G. and Katritzky, A.R. (2000). Aromaticity: A theoretical concept of immense practical importance. *Tetrahedron*. 56: 1783- 1796.

[55] Schleyer, P.R. (2001). Introduction: Aromaticity. *Chem. Rev*. 101: 1115-1118.

[56] Cranks, M.K. Krygowski, T.M. Katritzky, A.R. and Schleyer, P.R. (2002). To what extent can aromaticity be defined uniquely? *J Org. Chem*. 67: 1333-1338.

[57] Schleyer, P.R. Maerker, C. Dransfeld, A. Jiao, H. and Hommes, N.J.R.E. (1996). Nucleus-independent chemical shifts: A simple and efficient aromaticity probe. *J Am Chem. Soc.*118: 6317-6318.

[58] Jiao, H. and Schleyer, P.R. (1998). Aromaticity of pericyclic reaction transition structures: magnetic evidence. *J Phys Org Chem.* 11: 655-662.

[59] Schleyer, P.R. Kiran, B.Simion, D.V. andSorensen, T.S. (2000). Does $Cr(CO)_3$ complexation reduce the aromaticity of benzene? *J Am Chem. Soc.* 122:510-513.

[60] Quinonero, D., Garau, C., Frontera, A., Ballaster, P., Costa, A., Deya, P.M. (2002).Quantification of aromaticity in oxocarbons: The problem of the fictitious "nonaromatic" reference system. *Chem. Eur J.* 8: 433-438.

[61] Patchkovskii, S. and Thiel, W. (2002). Nucleus-independent chemical shifts from semiempirical calculations. *J Mol Model.* 6: 67-75.

[62] Stanger, A. (2006) Nucleus-independent chemical shifts (NICS): Distance dependence and revised criteria for aromaticity and antiaromaticity. *J Org Chem. 71* (3):883–893.

[63] Stanger, A. (2009). What is... aromaticity: a critique of the concept of aromaticity—can it really be defined? *Chem. Commun*: 1939-1947.

[64] Solà, M., Feixas, F., Jiménez-Halla, J.O.C., Matito, E. and Poater, J. (2010). A critical assessment of the performance of magnetic and electronic indices of aromaticity, *Symmetry*. 2: 1156-1179.

[65] Clar, E. (1972). *The aromatic sextet*. London, Wiley.

In: Advances in Chemistry Research
Editor: James C. Taylor
ISBN: 978-1-53615-828-1

Chapter 5

SOLUTIONS OF FORMALDEHYDE IN WATER AND ALCOHOLS

*Michael M. Silaev**
Faculty Chemistry, Lomonosov Moscow State University,
Vorobievy Gory, Moscow, Russia

ABSTRACT

A mechanism of the initiated nonbranched-chain process of forming 1,2-alkanediols and carbonyl compounds in alcohol–formaldehyde systems is suggested. The quasi-steady-state treatment is used to obtain kinetic equations that can describe the nonmonotonic (with a maximum) dependences of the formation rates of the products on the concentration of free unsolvated formaldehyde. The experimental concentration of the free unsolvated form of formaldehyde are given at the different temperatures, solvent permittivity, and total concentrations of formaldehyde in water and alcohols. An empirical equation for calculating the free formaldehyde concentration in alcohol–formaldehyde (including water/ethanediol–formaldehyde) systems at various temperatures and total formaldehyde concentrations and an equation for evaluating solvent concentrations in these systems were derived.

* Corresponding Author's E-mail: mmsilaev@rc.chem.msu.ru.

Keywords: nonbranched-chain process, free formaldehyde, 1-hydroxyalkyl radical, formyl radical, competing reaction, equation

INTRODUCTION

Free radicals add to the carbon atom at the double bond of the carbonyl group of dissolved free (unsolvated, monomer) formaldehyde. The concentration of free formaldehyde in the solution at room temperature is a fraction of a percent of the total formaldehyde concentration, which includes formaldehyde chemically bound to the solvent [1]. The concentration of free formaldehyde exponentially increases with increasing temperature [2]. The energy released as a result of this addition, when the C = O bond is converted into an ordinary bond, is 30 to 60 kJ mol^{-1} (according to the data on the addition of C_1–C_4 alkyl radicals in the gas phase under standard conditions [3–6]). The resulting free 1:1 adduct radicals can both abstract hydrogen atoms from the nearest-neighbor molecules of the solvent or unsolvated formaldehyde and, due to its structure, decompose by a monomolecular mechanism including isomerization [7, 8].

The analysis of stable liquid-phase products was carried out by the gas chromatographic method. The quasi-steady-state treatment is used to obtain kinetic equations.

ADDITION OF 1-HYDROXYALKLYL FREE RADICALS WITH TWO OR MORE CARBON ATOMS

Free 1-hydroxyalkyl radicals (which result from the abstraction of a hydrogen atom from the carbon atom bonded to the hydroxyl group in molecules of saturated aliphatic alcohols but methanol under the action of chemical initiators [9, 10], light [11, 12], or ionizing radiation [13, 14]) add at the double bond of free formaldehyde dissolved in the alcohol, forming 1,2-alkanediols [7–10, 12–18], carbonyl compounds, and methanol [14, 15] *via* the chaining mechanism. (The yields of the latter two products in the temperature range of 303 to 448 K are one order of magnitude lower). In these processes, the determining role in the reactivity of the alcohols can be played by the desolvation of formaldehyde in alcohol–formaldehyde solutions, which depends both on the temperature and on the polarity of the solvent [2, 14]. For the γ-radiolysis of 1(or 2)-propanol–formaldehyde system at a constant temperature, the dependences of the radiation-chemical yields of 1,2-alkanediols and carbonyl compounds as a function of the formaldehyde concentration show maxima and are symbatic [13, 15]. For a constant total formaldehyde concentration of 1 mol dm^{-3}, the dependence of the 1,2-alkanediol yields as a function of temperature for 303–473 K shows a maximum, whereas the yields of carbonyl compounds and methanol increase monotonically [14] (along with the concentration of free formaldehyde [2]). In addition to the above products, the nonchain mechanism in the γ-radiolysis of the solutions of formaldehyde in ethanol and 1- and 2-propanol gives ethanediol, carbon monoxide, and hydrogen in low radiation-chemical yields (which, however, exceed the yields of the same products in the γ-radiolysis of individual alcohols) [7, 14, 15]. The available experimental data can be described in terms of the following scheme of reactions.

In these reactions, I is an initiator, e.g., a peroxide [9, 10]; $R_0^\bullet$, some reactive radical (initiator radical); R, an alkyl; ROH, a saturated aliphatic alcohol, either primary or secondary, beginning from ethanol; CH_2O, the unsaturated molecule – free formaldehyde; $^\bullet CH_2OH$, the 1-hydroxymetyl fragment radical; $^\bullet R_{(-H)}OH$, the reactive 1-hydroxyalkyl addend radical, beginning from 1-hydroxyethyl; $R_{(-H)}(OH)CH_2O^\bullet$, the reactive hydroxyalkoxyl 1:1 adduct radical; $^\bullet CHO$, the low-reactive formyl radical (inhibitor radical); R_0H, the molecular product; $R_{(-H)}(OH)CH_2OH$, 1,2-alkanediol; $R_{(-2H)}HO$, an aldehyde in the case of a primary alcohol and an R'R"CO ketone in the case of a secondary alcohol; $R_{(-H)}(OH)R_{(-H)}OH$, a vicinal alkanediol; $R_{(-H)}(OH)CHO$, a hydroxyaldehyde. The chain evolution stage of *Scheme* includes consecutive reaction pairs 2–3, 2–3a, and 3a–3b; parallel (competing) reaction pairs 3–3a, 3–3b, 3–4, and 3a–4; and consecutive–parallel reactions 2 and 4.

Scheme

Chain initiation

1. $I \xrightarrow{2k_1} 2R_0^{\bullet}$;

1a. $R_0^{\bullet} + ROH \xrightarrow{k_{1a}} ROH + {}^{\bullet}R_{(-H)}OH$.

Chain propagation

2. ${}^{\bullet}R_{(-H)}OH + CH_2O \xrightarrow{k_2} R_{(-H)}(OH)CH_2O^{\bullet}$;

3. $R_{(-H)}(OH)CH_2O^{\bullet} + ROH \xrightarrow{k_3} R_{(-H)}(OH)CH_2OH + {}^{\bullet}R_{(-H)}OH$;

3a. $R_{(-H)}(OH)CH_2O^{\bullet} \xrightarrow{k_{3a}} R_{(-2H)}HO + {}^{\bullet}CH_2OH$
(or $R'R''CO + {}^{\bullet}CH_2OH$);

3b. ${}^{\bullet}CH_2OH + ROH \xrightarrow{k_{3b}} CH_3OH + {}^{\bullet}R_{(-H)}OH$.

Inhibition

4. $R_{(-H)}(OH)CH_2O^{\bullet} + CH_2O \xrightarrow{k_4} R_{(-H)}(OH)CH_2OH + {}^{\bullet}CHO$.

Chain termination

5. $2{}^{\bullet}R_{(-H)}OH \xrightarrow{2k_5} R_{(-H)}(OH)R_{(-H)}OH$
(or: $ROH + R_{(-2H)}HO$,
$ROH + R'R''CO$);

6. ${}^{\bullet}R_{(-H)}OH + {}^{\bullet}CHO \xrightarrow{k_6} R_{(-H)}(OH)CHO$
(or: $R_{(-2H)}HO + CH_2O$,
$R'R''CO + CH_2O$,
$ROH + CO$);

7. $2{}^{\bullet}CHO \xrightarrow{2k_7} HC(O)CHO$
(or: $CH_2O + CO$,
$2CO + H_2$).

In addition, it seems unlikely that free adduct radicals will add to formaldehyde at higher temperatures the reaction of adding is unlikely because this would result in an ether bond. The addition of hydroxymethyl radicals to formaldehyde, which is in competition with reaction 3b, is not included as well, because there is no chain formation of ethanediol at 303–448 K [14]. At the same time, small amounts of ethanediol can form *via* the dimerization of a small fraction of hydroxymethyl radicals, but this cannot have any appreciable effect on the overall process kinetics. The addition of free formyl radicals to formaldehyde cannot proceed at a significant rate, as is indicated by the fact that there is no chain formation of glycol aldehyde in the systems examined [14].

The mechanism of the decomposition of the free adduct radical *via* reaction 3a, which includes the formation of an intramolecular H···O bond and isomerization, can be represented as follows [7, 8, 15]:

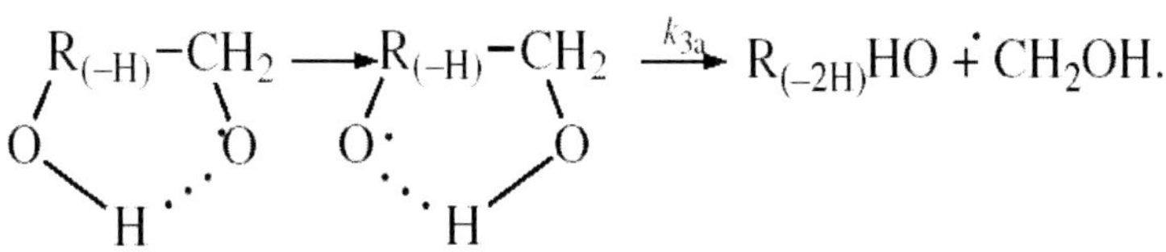

The probability of the occurrence of reaction 3a should increase with increasing temperature. This is indicated by experimental data presented above [7, 8, 15]. The decomposition of the hydroxyalkoxyl radical. $R_{(-H)}(OH)CH_2O^{\bullet}$ (reaction 3a) is likely endothermic. The endothermic nature of reaction 3a is indirectly indicated by the fact that the decomposition of simple C_2–C_4 alkoxyl radicals $RO^{\bullet}$ in the gas phase is accompanied by heat absorption: (ΔH°_{298} = 30–90 kJ mol^{-1} [4–6]). Reaction 3b, subsequent to reaction 3a, is exothermic, and its heat for C_2–C_3 alcohols in the gas phase is ΔH°_{298} = –40 to –60 kJ mol^{-1} [4–6]. As follows from the above scheme of the process, reactions 3a and 3b, in which the formation and consumption of the highly reactive free radical hydroxymethyl take place (at equal rates under steady-state conditions), can be represented as a single bimolecular reaction 3a,b occurring in a "cage" of solvent molecules.

The free formyl radical resulting from reaction 4, which is in competition with reactions 3 and 3a, is comparatively low-reactive because its spin density can be partially delocalized from the carbon atom *via* the double bond toward the oxygen atom, which possesses a higher electron affinity [3]. For example, in contrast to the methyl and alkoxyl π-radicals, the formyl σ-radical can be stabilized in glassy alcohols at 77 K [19]. In the gas phase, the dissociation energy of the C–H bond in formyl radicals is half that for acetyl radicals and is about 5 times lower than the dissociation energy of the C_α–H bond in saturated C_1–C_3 alcohols [3].

As distinct from reactions 3 and 3a,b, reaction 4 leads to an inefficient consumption of hydroxyalkoxyl adduct radicals, without regenerating the initial 1-hydroxyalkyl addend radicals. Reaction 4 together with reaction 6 (mutual annihilation of free formyl and chain-carrier 1-hydroxyalkyl radicals) causes the inhibition of the nonbranched-chain process. For the disproportionation of the free radicals, the heats of reactions 5–7 for C_1–C_3 alcohols in the gas phase vary in the range of $\Delta H^{\circ}_{298} = -135$ to -385 kJ mol^{-1} [3–6].

The rates of the chain formation of 1,2-alkanediols in reaction 3 (and their nonchain formation in reaction 4), carbonyl compounds in reaction 3a, and methanol in reaction 3b are given by the following equations[1]:

$$V_{3,4}(R_{(-H)}(OH)CH_2OH) = \frac{V_1(\alpha l + x)k_2 x}{k_2 x^2 + (\alpha l + \beta + x)\sqrt{2k_5 V_1}}, \quad (1)$$

$$V_{3a}(R_{(-2H)}HO) = V_{3b}(CH_3OH) = \frac{V_1 \beta k_2 x}{k_2 x^2 + (\alpha l + \beta + x)\sqrt{2k_5 V_1}}, \quad (2)$$

where V_1 is the initiation rate, l is the molar concentration of the saturated alcohol at the given total concentration c_0 of formaldehyde[2] dissolved in it,

---

[1] The rate equations were derived by quasi-steady-state treatment, which is most suitable for describing the processes including at least eight to ten reactions with four to six different free radicals and at most three to seven experimental points in their functional curves, using the condition for the first steps of the process that makes it possible to reduce the exponent of term $2k_5[^{\bullet}R_{(-H)}OH]^2$ to 1 in equation $d[R^{\bullet}]/dt = 0$ [7, 15]: $k_6 = \sqrt{2k_5 2k_7}$ [20] and $V_1 = V_5 + 2V_6 + V_7 = (\sqrt{2k_5}[^{\bullet}R_{(-H)}OH] + \sqrt{2k_7}[^{\bullet}CHO])^2$. The relations between the reaction rates and radiation-chemical yields for radiation-chemical processes are reported by Silaev [21].

[2] The alcohol concentration in alcohol–formaldehyde solutions at any temperature can be estimated by the method suggested in [22, 23]. The data necessary for estimating the concentration of free

x is the molar concentration of free formaldehyde ($l >> x$), k_2 is the rate constant of reaction 2 (addition of 1-hydroxyalkyl free radical to free formaldehyde), and $\alpha = k_3/k_4$ and $\beta = k_{3a}/k_4$ (mol dm^{-3}) are the ratios of the rate constants of the competing (parallel) reactions. Estimates of $2k_5$ were reported by Silaev *et al.* [22, 24]. From the extremum condition for the reaction 3a rate function, $\partial V_{3a} / \partial x = 0$, we derived the following analytical expression:

$$k_2 = (\alpha l_m + \beta)\sqrt{2k_5V_1} / x_m^2 .$$

Equations (1) and (2) taken on the condition that $k_2x^2 >> (\alpha l + x)\sqrt{2k_5V_1}$ (the descending branch of the curve having a maximum) can be transformed to equations that can be used for a preliminary estimation of parameters α and β:

$$V_{3,4} = (V_1/\varphi)[(\alpha l/x) + 1], \tag{1a}$$

$$V_{3a} = V_{3b} = V_1\, \beta/\varphi\, x, \tag{2a}$$

where $\varphi = 2$ for the maximum (when $k_2x^2 \cong (\alpha l + x)\sqrt{2k_5V_1}$) and $\varphi = 1$ for the descending portion of the curve.

The overall process rate is a complicated function of the formation and disappearance rates of the $^{\bullet}R_{(-H)}OH$ and $^{\bullet}CHO$ free radicals:

$$V(R_{(-H)}(OH)CH_2OH,\ R_{(-2H)}HO,\ CH_3OH) =$$
$$V_{1a} + V_3 + V_{3b} - V_4 - V_5 + V_7. \tag{3}$$

The ratios of the rates of the competing reactions are $V_3/V_4 = \alpha l/x$ and $V_{3a}/V_4 = \beta/x$, and the chain length is $\nu = (V_3 + V_{3a})/V_1$. The ratio of the rates

formaldehyde using the total formaldehyde concentration in the solution are reported by Silaev *et al.* [2, 22].

of formation of 1,2-alkanediol and the carbonyl compound is a simple linear function of x:

$$V_{3,4}(R_{(-H)}(OH)CH_2OH)/V_{3a}(R_{(-2H)}HO) =$$
$$(k_4/k_{3a})\,x + (k_3/k_{3a})l. \quad (4)$$

The equations for the rates of chain-termination reactions 5–7 are given by the following equations:

$$V_5 = V_1^2 2k_5\left(\alpha l+\beta+x\right)^2/f^2, \quad (5)$$

$$2V_6 = 2V_1\sqrt{2k_5V_1}\left(\alpha l+\beta+x\right)k_2x^2/f^2, \quad (6)$$

$$V_7 = V_1\left(k_2x^2\right)^2/f^2, \quad (7)$$

where $f = k_2x^2 + (\alpha l + \beta + x)\sqrt{2k_5V_1}$.

Neutral formaldehyde solutions in alcohols at room temperature primarily consist of a mixture of formaldehyde polymer solvates reversibly bound to alcohols; these polymer solvates differ in molecular mass and have the general formula $RO(CH_2O)_nH$, where n = 1–4 [1]. The concentration of formaldehyde that occurs in solution as a free, unsolvated active species chemically unbound to the solvent (this species is capable of scavenging free radicals) at room temperature is lower than a percent of the total formaldehyde concentration [1]. The concentration x of the free formaldehyde species in solutions was determined by high-temperature UV spectrophotometry in the range 335–438 K at the total formaldehyde concentration c_0 (free and bound species including the concentration of polymer solvates) of 1.0–8.4 mol dm^{-3} in water, ethanediol, methanol, ethanol, 1-propanol, 2-propanol, and 2-methyl-2-propanol [2] (see Table of the Appendix). This concentration increases with temperature according to an exponential law, and it can be as high as a few percent of the total concentration in solution under the test conditions, up to 19.3% in the case of 2-methyl-2-propanol at a total concentration of 1.0 mol dm^{-3} and a temperature of 398 K. The following empirical equation relating the concentration x (mol dm^{-3}) of free formaldehyde to temperature T (K) and the total concentration c_0 in the solution (measured at room temperature), was developed by the treatment of 101 data points [2, 22]:

Table 1. Coefficients of the empirical Eq. (8) for the estimation of the concentration *x* of free formaldehyde in polar solvent–formaldehyde systems

Solvent	Coefficient		
	a	*b*	*h*
Water	2.36	4.45	0.80
Ethanediol	1.83	2.60	1.28
Methanol	3.11	5.58	0.22 $c_0/\lg c_0$
Ethanol	3.10	5.92	1.10 $(10^3/T) - 1.44$
1-Propanol	2.42	4.47	1.30
2-Propanol	2.42	4.64	1.05
2-Methyl-2-propanol	3.19	7.31	0.96

$$\lg x = -a\,(10^3/T) + b + h \lg c_0\,, \tag{8}$$

where the coefficients a and b were calculated as the parameters of a straight-line equation by the least-squares technique from the dependence of $\lg x$ on $1/T$ at c_0 = 1.0 mol dm^{-3} for various solvents, and the coefficient h was obtained as the average value of the slopes of $\lg x$ as linear functions of $\lg c_0$ at various series of fixed temperatures. The Table 1 summarizes these coefficients for each solvent. As regards the experimental data, the error in the calculations of the concentration x of free formaldehyde made by Eq. (8) in the specified temperature range was no higher than 25%.

On the assumption that the dependence of the density of a given solution on the concentration of formaldehyde is similar to the analogous linear dependence found for aqueous formaldehyde solutions (0–14 mol dm^{-3}; 291 K) [25], the concentrations l_T (mol dm^{-3}) of alcohols in alcohol–

formaldehyde solutions at a certain temperature can be estimated by the equation

$$l_T = \frac{(10^3 d - 21.6\, c_0)d_T}{(d + 8.4 \times 10^{-3}\, c_0\,)M}, \qquad (9)$$

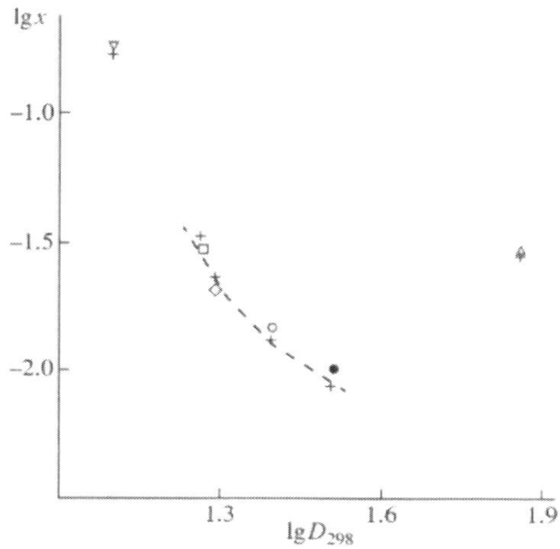

Figure 1. Logarithmic plot of the experimental concentrations x (mol dm^{-3}) of free formaldehyde at its total concentration c_0 = 1.0 mol dm^{-3} and 395 ± 3 K in (△) water, (●) methanol, (○) ethanol, (◇) 1-propanol, (□) 2-propanol, and (▽) 2-methyl-2-propanol as functions of the permittivity D_{298} of these solvents (+ refers to the concentrations x in the above solvents calculated by empirical Eq. (8)).

where c_0 is the total formaldehyde concentration (mol dm^{-3}); M is the molecular mass (g mol^{-1}) of the solvent; d and d_T are the solvent densities (g cm^{-3}) at room and given temperatures, respectively; the coefficients 8.4×10^{-3} and 21.6 have the units of 10^3g mol^{-1} and g mol^{-1}, respectively.

Earlier [2], it was found that the concentration x of the free formaldehyde species decreased with the solvent permittivity D_{298} at a constant temperature. Water is an exception. Although water is more polar than alcohols, the concentration x of free formaldehyde in an aqueous solution is anomalously high and reaches the level of its concentration in 2-propanol, all other factors being the same (see Figure 1) [2, 22]. This can be due to the specific instability of hydrated formaldehyde species and the ease of their conversion into free formaldehyde with increasing temperature.

As an example, the Figure 2 illustrates the use of Eqs. (1) and (2) for describing the experimental dependences of the formation rates of 1,2-butanediol (curve *1*) in reactions 3 and 4 and propanal (curve *2*) in reaction 3a on the concentration x of free formaldehyde in the 1-propanol–formaldehyde reacting system at the total formaldehyde concentration c_0 of 2.0 to 9.5 mol dm^{-3} and temperature of 413 K [7, 15, 26]. The concentration dependence of the propanal formation rate was described using the estimates of kinetic parameters obtained for the same dependence of the 1,2-butanediol formation rate. We considered these data more reliable for the reason that the carbonyl compounds forming in the alcohol–formaldehyde systems can react with the alcohol and this reaction depends considerably on the temperature and acidity of the medium [1].

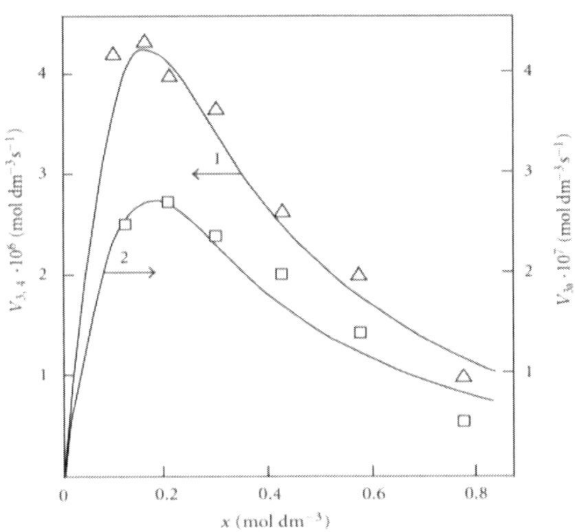

Figure 2. Reconstruction of the functional dependence (curves) of the product formation rates $V_{3,4}$ and V_{3a} on the concentration x of free formaldehyde (model optimization with respect to the parameters α, β and k_2) from empirical data (symbols) for the 1-propanol–formaldehyde system at 413 K [7, 15, 26]: (*1*, △) calculation using Eq. (1), the product is 1,2-butanediol, standard deviation of $S_Y = 2.20 \times 10^{-7}$; (*2*, □) calculation using Eq. (2), the product is propanal, $S_Y = 2.38 \times 10^{-8}$.

The mathematical modeling of the process was carried out using a ^{137}Cs γ-radiation dose rate of P = 0.8 Gy s^{-1} [13, 26], a total initiation yield of $G(CH_3CH_2\dot{C}HOH)$ = 9.0 particles per 100 eV [7, 15] ($V_1 = 4.07 \times 10^{-7}$ mol dm^{-3} s^{-1}), and $2k_5 = 4.7 \times 10^9$ dm^3 mol^{-1} s^{-1}. The following values of the parameters were obtained: $\alpha = 0.36 \pm 0.07$, $\beta = 0.25 \pm 0.05$ mol dm^{-3}, and $k_2 = (6.0 \pm 1.4) \times 10^3$ dm^3 mol^{-1} s^{-1}. Note that, as compared to the yields of 1,2-propanediol in the γ-radiolysis of the ethanol–formaldehyde system, the yields of 2,3-butanediol in the γ-radiolysis of the ethanol–acetaldehyde system are one order of magnitude lower [26]. Using data from [7, 15], it can be demonstrated that, at 433 K, the double bond of 2-propen-1-ol accepts the 1-hydroxyethyl radical 3.4 times more efficiently than the double bond of formaldehyde [27].

ADDITION OF HYDROXYMETHYL FREE RADICALS

The addition of hydroxymethyl free radicals to the carbon atom at the double bond of unsolvated formaldehyde molecules in methanol results in the chain formation of ethanediol [16]. In this case, reaction 3a in *Scheme* is the reverse of reaction 2, the 1-hydroxyalkyl radical $^{\bullet}R_{(-H)}OH$ is the hydroxymethyl radical $^{\bullet}CH_2OH$, so reaction 3b is eliminated ($k_{3b} = 0$), and reaction 5 yields an additional amount of ethanediol *via* the dimerization of chain-carrier hydroxymethyl radicals (their disproportionation can practically be ignored [28]). The scheme of these reactions is presented in [17]. The experimental dependence of the ethanediol radiation-chemical yields on formaldehyde concentration in γ-radiolysis of methanol–formaldehyde system at 373–473 K is shown in Figure 3. The maximum yield of the product reached 139 molecule per 100 eV at a formaldehyde concentration of 1.4 mol dm^{-3} in methanol and a temperature of 473 K. The dependence of the yield on the concentration at 373 K does not have a maximum even at the highest (3.1 mol dm^{-3}) concentration of formaldehyde, whereas maximums appear on the curves in the formaldehyde concentration region about 1.4 mol dm^{-3} at higher temperatures.

The rate equation for ethanediol formation by the chain mechanism in reaction 3 and by the nonchain mechanism in reactions 4 and 5 in the methanol–formaldehyde system has a complicated form[3] as compared to Eq. (1) for the formation rate of the other 1,2-alkanediols [8]:

$$V_{3,4,5}(CH_2OH)_2 = V_1[f(\alpha l + x)k_2x + V_1 2k_5(\alpha l + \beta + x)^2]/f^2\,, \qquad (10)$$

where $f = k_2x^2 + (\alpha l + \beta + x)\sqrt{2k_5V_1}$.

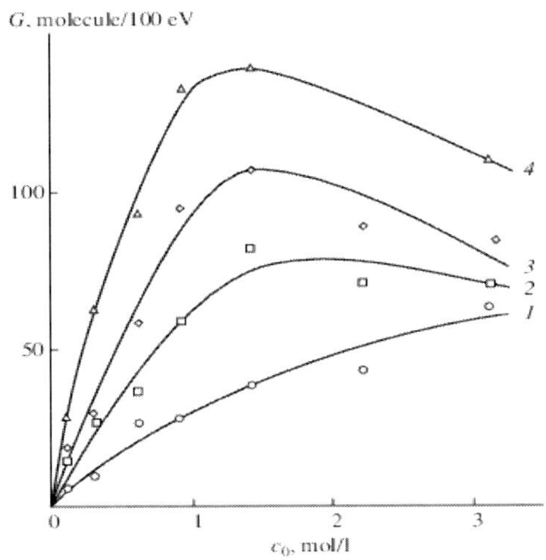

Figure 3. Dependence of the ethanediol yield G on the total concentration c_0 of formaldehyde upon γ-radiolysis of the methanol–formaldehyde system at temperatures of (*1*) 373, (*2*) 423, (*3*) 448, and (*4*) 473 K [16].

Equation (10) when $k_2x^2 \ll (\alpha l + \beta + x)\sqrt{2k_5V_1}$ (ascending branch of the curve having a maximum) and $\alpha l \gg \beta$ (practically without reaction 3a) is transformed to a simple directly-proportional dependence on the concentration x of free formaldehyde, which can be used to pre-estimate the parameter k_2:

[3]In an earlier publication [15], this equation does not take into account reaction 3a.

$$V_{3,4,5} = (V_1/\varphi^2)[(\varphi\, k_2\, x/\sqrt{2k_5V_1}\,) + 1], \tag{10a}$$

where $\varphi = 1$ for the ascending portion of the curve and $\varphi = 2$ for the maximum, when $k_2x^2 \cong (\alpha l + \beta + x)\sqrt{2k_5V_1}$.

If the rate of ethanediol formation by the dimerization mechanism in reaction 5 is ignored for the reason that it is small as compared to the total rate of ethanediol formation in reactions 3 and 4, Eq. (10) will be identical to Eq. (1). After the numerator and denominator on the right-hand side of Eq. (1) are divided by $k_{-2} \equiv k_{3a}$, one can replace k_2 in this equation with $K_2 = k_2/k_{-2}$, which is the equilibrium constant for the reverse of reaction 2. Ignoring the reverse step of reaction 2 ($k_{3a} = 0$, $\beta = 0$) can further simplify Eq. (1). In this case, the rate constant k_2 is effective.

CONCLUSION

In summary, the material on the kinetics of nonbranched-chain addition of free saturated 1-hydroxyalkyl radicals to the double bonds of free formaldehyde molecules makes it possible to describe, using Eq. (1) obtained by quasi-steady-state treatment, experimental dependences with a maximum of the formation rates of molecular 1:1 adducts (1,2-alkanediols) on the concentration of an unsaturated compound over the entire region of its change in binary reaction systems consisting of saturated (alcohol) and unsaturated (free formaldehyde) components (Figure 2).

The proposed addition mechanism involves the reaction of a free hydroxyalkoxyl 1:1 adduct radical with an unsaturated free formaldehyde molecule yielding a low-reactive free formyl radical (the reaction 4 competing with the chain propagation reactions in the *Scheme*). In such binary reaction systems, the unsaturated compound is both a reactant and an autoinhibitor, specifically, a source of low-reactive free radicals shortening kinetic chains.

References

[1] Ogorodnikov, S. K. *Formal'degid (Formaldehyde),* Khimiya, Leningrad, 1984.

[2] Silaev, M. M., A. V. Rudnev, and E. P. Kalyazin, "Formaldehyde. III. Concentration of Free Formaldehyde as a Function of Temperature, Polarity of Solvents, and Total Concentration of Formaldehyde in Solution", *Zhurnal Fizicheskoi Khimii*, 1979, vol. 53, no. 7, pp. 1647–1651.

[3] Gurvich, L. V., G. V. Karachevtsev, V. N. Kondrat'ev, Yu. A. Lebedev, V. A. Medvedev, V. K. Potapov, and Yu. S. Khodeev, *Energii razryva khimicheskikh svyazei. Potentsialy ionizatsii i srodstvo k elektronu* [*Bond Dissociation Energies, Ionization Potentials, and Electron Affinity*], V. N. Kondrat'ev, Editor, Nauka, Moscow, 1974.

[4] Benson, S. W. *Thermochemical Kinetics: Methods for the Estimation of Thermochemical Data and Rate Parameters,* 2nd Edition, Wiley, New York, 1976.

[5] Pedley, J. B., R. D. Naylor, and S. P. Kirby, *Thermochemical Data of Organic Compounds,* 2nd Edition, Chapman & Hall, London, 1986.

[6] Orlov, Yu. D., Yu. A. Lebedev, and I. Sh. Saifullin, *Termokhimiya organicheskikh svobodnykh radikalov* [*Thermochemistry of Organic Free Radicals*], A. M. Kutepov, Editor, Nauka, Moscow, 2001.

[7] Silaev M. M. and L. T. Bugaenko, "Kinetics of the Addition of α-Hydroxyalkyl Radicals to 2-Propen-1-ol and Formaldehyde", *Kinetics and Katalysis*, 1994, vol. 35, no. 4, pp. 509–513, English Translation in: *Kinetics and Katalysis*, 1994, vol. 35, no. 4, 1994, pp. 463–467.

[8] Silaev, M. M. "Simulation of Nonbranched Chain Processes for Producing 1,2-Alkanediols in Alcohol–Formaldehyde Systems", *Teoreticheskie Osnovy Khimicheskoi Tekhnologii*, 2007, vol. 41, no. 4, pp. 379–384, English Translation in: *Thoretical Foundations Chemical Engineering*, 2007, vol. 41, no. 4, pp. 357–361.

[9] Oyama, M., "A Free-Radical Reaction of Primary and Secondary Alcohols with Formaldehyde", *The Journal of Organic Chemistry*, 1965, vol. 30, no. 7, pp. 2429–2432.

[10] Nikishin, G. I., D. Lefor, and E. D. Vorob'ev, "Free Radical Reaction of Primary Alcohols with Formaldehyde", *Izvestiya Akademii Nauk SSSR, Ser. Khimiya*, 1966, no. 7, pp. 1271–1272.

[11] Urry, W. H., F. W. Stacey, E. S. Huyser, and O. O. Juveland, "The Peroxide- and Light-Induced Additions of Alcohols to Olefins", *Journal of the American Chemical Society*, 1954, vol. 76, no. 2, pp. 450–455.

[12] Dzhurinskaya, M. B., A. V. Rudnev, and E. P. Kalyazin, "High Temperature UV Photolysis of Formaldehyde in Liquid Methanol", *Vestnik Moskovskogo Universiteta, Ser. 2: Khimiya*, 1984, vol. 25, no. 2, pp. 173–176.

[13] Kalyazin, E. P., E. P. Petryaev, and O. I Shadyro, "Reaction between Oxyalkyl Radicals and Aldehydes", *Zhurnal Organicheskoi Khimii*, 1977, vol. 13, no. 2, pp. 293–295.

[14] Novoselov, A. I., A.I., Silaev, M.M., and L. T. Bugaenko, "Effect of Temperature on the Yields of Final Products in the γ-Radiolysis of Formaldehyde Solutions in C_1–C_3 Alkanols", *Khimiya Vysokich Energii*, 2004, vol. 38, no. 4, pp. 270–272, English Translation in: *High Energy Chemistry*, 2004, vol. 38, no. 4, 2004, pp. 236–238.

[15] Silaev M. M. and L. T. Bugaenko, "Mathematical Simulation of the Kinetics of Radiation Induced Hydroxyalkylation of Aliphatic Saturated Alcohols", *Radiation Physics and Chemistry*, 1992, vol. 40, no. 1, pp. 1–10.

[16] Novoselov, A. I., M. M. Silaev, and L. T. Bugaenko, "Dependence of Ethanediol Yield on Formaldehyde Concentration in γ-Radiolysis of Methanol–Formaldehyde System at 373–473 K", *Khimiya Vysokikh Energii*, 2008, vol. 42, no. 1, pp. 74–75, English Translation in: *High Energy Chemistry*, 2008, vol. 42, no. 1, pp. 69–70.

[17] Novoselov, A. I., M. M. Silaev, and L. T. Bugaenko, "γ-Induced Single-Step Synthesis of Ethylene Glycol from Methanol–Formaldehyde Solution", *Theoreticheskie Osnovy Khimicheskoy Tekhnologii*, 2010, vol. 44, no. 4, pp. 450–453, English Translation in: *Theoretical Foundation of Chemical Engineering*, 2010, vol. 44, no. 4, pp. 432–435.

[18] Novoselov, A. I., M. M. Silaev, and L. T. Bugaenko, "Dependence of 1,2-Propanediol Yield on Formaldehyde Concentration in γ-Radiolysis of Ethanol–Formaldehyde System at 373–473 K", *Khimiya Vysokikh Energii*, 2007, vol. 41, no. 1, p. 58, English Translation in: *High Energy Chemistry*, 2007, vol. 41, no. 1, p. 53.

[19] Pshezhetskii, S. Ya., A. G. Kotov, V. K. Milinchuk, V. A. Roginskii, and V. I. Tupikov, *EPR svobodnykh radikalov v radiatsionnoi khimii* [*ESR of Free Radicals in Radiation Chemistry*], Khimiya, Moscow, 1972.

[20] Bateman, L. "Olefin Oxidation", *Quarterly Reviews*, 1954, vol. 8, no. 2, pp. 147–167.

[21] Silaev, M. M. "Simulation of the Nonbranched-Chain Addition of Saturated Free Radicals to Alkenes and Their Derivatives Yielding 1:1 Adducts", *Teoreticheskie Osnovy Khimicheskoi Tekhnologii*, 2007, vol. 41, no. 3, pp. 280–295, English Translation in: *Theoretical Foundations of Chemical Engineering*, 2007, vol. 41, no. 3, pp. 273–278.

[22] Silaev, M. M. "Applied Aspects of the γ-Radiolysis of C_1–C_4 Alcohols and Binary Mixtures on Their Basis", *Khimiya Vysokikh Energii*, 2002, vol. 36, no. 2, pp. 97–101, English Translation in: *High Energy Chemistry*, 2002, vol. 36, no. 2, pp. 70–74.

[23] Silaev, M. M. "Estimating the Solvent Concentration in Formaldehyde Solutions at Various Temperatures", *Zhurnal Fizicheskoy Khimii*, 1993, vol. 67, no. 9, p. 1944.

[24] Silaev, M. M., L. T. Bugaenko, and E. P. Kalyazin, "On the Possibility of Adequately Estimating of the Rate Constants for the Reaction of Hydroxyalkyl Radicals with Each Other Using the Self-Diffusion Coefficients or Viscosities of the Corresponding Alcohols", *Vestnik*

Moskovskogo. Univiversiteta, Ser. 2: Khimiya, 1986, vol. 27, no. 4, pp. 386–389.

[25] Walker, J. F. *Formaldehyde*, Reinhold, New York, 1953, English Translation under the title Formal'degid, Goskhimizdat, Moscow, 1957, p. 106.

[26] Shadyro, O. I. *Radiation-chemical Conversions of Aldehydes in Various Systems*, PhD Thesis (Chemistry), Belarusian State University, Minsk, 1975.

[27] Silaev, M. M. "Relative Reactivity of α-Hydroxyethyl Radicals for 2-Propene-1-ol and Formaldehyde Double-Bond Addition", *Vestnik Moskovskogo Universiteta, Ser. 2: Khimiya*, 1993, vol. 34, no. 3, p. 311.

[28] Seki, H., R. Nagai, and M. Imamura, "γ-Radiolysis of a Binary Mixture of Methanol and Water. The Formation of Formaldehyde in the Radiolysis of Liquid Methanol", *Bulletin of the Chemical Society of Japan*, 1968, vol. 41, no. 12, pp. 2877–2881.

APPENDIX

Table A-1. The experimental concentrations x (mol dm^{-3}) of free formaldehyde at different temperatures T (K) and total formaldehyde concentrations c_0 (mol dm^{-3}) in various solvents

c_0	T	10^2x	c_0	T	10^2x	c_0	T	10^2x	c_0	T	10^2x
Water			4.44	389	5.20	4.0	381	5.00	1.8	371	2.08
1.0	358	0.78	4.44	405	7.50	4.0	397	8.80	1.8	393	6.00
1.0	387	2.22	4.44	418	10.0	4.0	409	12.00	1.8	418	12.20
1.0	393	**3.23**	**Methanol**			6.2	347	2.80	1.8	438	16.70
1.0	407	4.55	1.0	375	0.33	6.2	376	7.80	3.0	343	1.25
2.0	353	1.44	1.0	395	1.00	6.2	393	12.50	3.0	375	5.40
2.0	387	4.70	1.0	423	2.90	**1-Propanol**			3.0	403	15.80
2.0	397	6.60	2.5	373	0.60	1.0	371	0.83	3.0	413	19.40
2.0	407	8.55	2.5	385	1.15	1.0	393	2.10	5.6	343	2.80

c_0	T	10^2x	c_0	T	10^2x	c_0	T	10^2x	c_0	T	10^2x
4.0	343	0.78	2.5	398	1.80	1.0	413	4.30	5.6	358	3.35
4.0	363	2.33	5.4	351	0.78	1.0	435	7.65	5.6	363	5.80
4.0	385	6.45	5.4	383	3.70	1.9	353	0.70	5.6	371	6.50
4.0	403	8.90	5.4	398	6.80	1.9	383	3.06	5.6	383	12.10
4.0	413	11.10	7.0	365	4.70	1.9	405	7.65	**2-Methyl-2-propanol**		
6.0	351	2.22	7.0	383	12.50	1.9	417	11.70	1.0	347	1.20
6.0	375	6.70	7.0	391	16.00	4.0	349	1.67	1.0	367	4.50
6.0	389	10.70	**Ethanol**			4.0	373	6.10	1.0	387	11.00
6.0	398	14.10	1.0	367	0.33	4.0	393	13.30	1.0	398	19.30
8.4	364	5.50	1.0	387	0.67	6.0	338	1.39	2.0	335	1.10
8.4	376	8.32	1.0	397	1.45	6.0	357	5.00	2.0	357	4.30
8.4	388	10.97	1.0	413	2.70	6.0	377	11.70	2.0	375	13.00
Ethanediol			1.0	423	4.00	6.0	389	18.30	2.0	383	18.50
1.0	409	1.30	2.0	373	1.10	7.8	343	3.06	3.0	338	1.70
1.0	418	1.80	2.0	394	2.90	7.8	358	6.25	3.0	353	4.70
1.0	435	2.45	2.0	409	5.80	7.8	377	16.90	3.0	365	9.60
3.33	358	1.20	2.0	419	8.20	**2-Propanol**			3.0	373	15.50
3.33	387	3.30	3.0	361	1.20	1.0	365	0.98	6.0	345	6.90
3.33	401	5.10	3.0	387	3.70	1.0	393	3.05	6.0	351	9.00
3.33	415	7.20	3.0	409	7.80	1.0	411	6.00	6.0	361	13.40
4.44	338	1.00	4.0	355	2.30	1.0	433	10.40	6.0	365	18.30

In: Advances in Chemistry Research
Editor: James C. Taylor
ISBN: 978-1-53615-828-1

Chapter 6

APPLICATION OF LUMINOL AS SOLVATOCHROMIC PROBE AND FLUORESCENT BIOMARKER FOR HETEROGENEOUS MEDIA

N. Shaemningwar Moyon, Imocha Rajkumar Singh, Mostofa Ataur Rohman, Prayasee Baruah, Vikash Kumar Sonu and Sivaprasad Mitra*

Center for Advanced Studies, Department of Chemistry, North-Eastern Hill University, Shillong, India

ABSTRACT

Luminol (5-amino-2,3-dihydro-1,4-phthalazinedione) is a versatile chemical that shows striking blue chemiluminescence in presence of certain metal ions when treated with appropriate oxidizing agent like hydrogen peroxide. This unique feature of luminol is often exploited by forensic investigators to detect trace amount of blood left in the crime

* Corresponding Author's E-mail: smitra@nehu.ac.in; smitranehu@gmail.com.

scene. The chemiluminescence behavior of luminol has been extensively used by the biologists as a cellular assay to detect copper, iron and cyanides etc. It has also been applied to quantify the oxygen secretion by phagocytizing cells and detection of proteins, cancer biomarkers etc. An important aspect of luminol chemiluminescence is its varying degree of sensitivity from one substance to another. For example, it is very sensitive to animal or human blood, organic tissues and fluids in comparison with other compounds containing metal ions such as paints, metallic surfaces, household products, or vegetable enzymes. The proven criminological and clinical application of luminol demands more extensive research work on it in the areas of photochemistry, fluorescence spectroscopy and biophysical chemistry.

A further point of interest is the possibility of using luminol as fluorescent probe, as it has been predicted to have hydrogen bonding characteristics akin to several biological entities. It may act as efficient fluorescent probe because of its sharp and strong absorption, as well as relatively high fluorescence quantum yield. Studying photophysical properties of model nitrogen heterocycles like luminol in homogeneous medium can lead to the understanding the comparative importance of solvatochromic parameters like α, β and π^* etc. on measurable spectroscopic quantities. The same processes investigated in the confined geometries ranging from "spheres" (micelle) and "cones" (cyclodextrin) to the more complex three-dimensional networks of proteins, DNA etc., would provide a unique opportunity that can lead to better understanding on the mechanism of drug recognition and action.

In this contribution, we will exploit the intense fluorescence behavior of luminol for quantitative understanding of hydrogen bonding behavior and its interaction with several biological media with the help of steady state and time-resolved measurements as well as density functional theory calculations.

Keywords: luminol, fluorescence, hydrogen bonding, DFT calculation, bio-molecular interaction

1. INTRODUCTION

Luminol (5-amino-2,3-dihydro-1,4-phthalazinedione, LH_2) is a versatile chemical which, when treated with appropriate oxidizing agent like hydrogen peroxide, shows striking blue chemiluminescence (CL) in presence of certain metal ions. This unique feature of LH_2 is often exploited by forensic investigators to detect trace amount of blood left in the crime scene and also applied as an aerosol in a mixture with sodium perborate, sodium carbonate and distilled water (Kricka 1995, Yeshion 1991). LH_2 is also used by the biologists as a cellular assay to detect several species like metal ions, hydrogen peroxide, nitrate, some alcohols, amines, amino acids, carbohydrates, cyanides, enzyme substrates and vitamins (Bowie, Sanders, and Worsfold 1996, Budavari 1996, Wu et al. 2008, Huntress, Stanley, and Parker 1934, Takayuki et al. 1986, Hoshino and Hinze 1987, Yeshion 1991, Seitz and Rudolf 1975). It has found wide application in different fields like forensic science, biological and curative science, quantification and detection of macro and micro molecules such as proteins, carbohydrates, DNA, and RNA (Barni et al. 2007, Gross, Harris, and Kaldun 1999, Li et al. 2010, Liu and Li 2011, Tan and Song 2014, Khan et al. 2014). Further, it has also been applied in environmental monitoring as biosensors, in the pharmaceutical industry for cellular localization and as biological tracers, in reporter gene-based assays and several other immunoassays (Karube et al. 1998, Wendler et al. 2005, Stockwell, Haggarty, and Schreiber 1999, Wang et al. 2011). Further, LH_2 enhanced CL probes have been used to characterize and quantify the secretion of oxygen by phagocytozing cells (Kahl et al. 1987). The use of LH_2 chemiluminescence has also been reported for facile detection of proteins (Bi, Zhou, and Zhang 2009), cancer biomarkers (Bi, Zhou, and Zhang 2009), as well as for reactive oxygen species produced by human neutrophils (Shleev et al. 2008). The ultra-high sensitivity of time-resolved chemiluminescence behavior of LH_2 can be used to measure OH/O_2^- radical species concentration as low as 0.2 μM in water (Wasselin-Trupin et al. 2000). An important aspect of LH_2 chemiluminescence is its different degree of sensitivity from one substance to another. LH_2 shows higher sensitivity to animal or human blood, organic tissues and fluids than to other compounds containing metal ions such as paints, metallic surfaces, household products, or vegetable enzymes. Therefore, the light emitted by LH_2 has different intensities and time duration depending on the material of contact and makes it an efficient forensic marker. Enhanced chemiluminescence has also been used to investigate the role of granulocyte-derived reactive oxygen species in heart muscle damage, to monitor polymorphonuclear leukocyte function in patients with diabetes mellitus, the activation of leukocytes in patients with peritonitis, the release of interleukin-8/interleukin-6 and the tumor necrosis factor-α from granulocytes following exposure to respiratory viral particles, and the effect of exposure to sulfur dioxide and sulfite aerosols on neutrophil function (Kricka 1995). Although, there is no evidence of LH_2 being currently used as a therapeutic agent, early clinical setting evidences for its use in the treatment of fluid accumulation of tissues in critically ill patients (Irie 1960b) as well as patches of baldness caused by auto-immune disease alopecia (Irie 1960c), promoting activity in blood clotting and wound healing (Irie 1960a, Lasovský and Grambal 1986) are found in literature. Now-a-days, LH_2 chemiluminescence in water is mostly applied in analytical purposes; for example, in special forensic medicine to detect trace amounts of blood. Therefore, LH_2 reactions in water have been intensely investigated. Solution phase spectroscopic properties of LH_2 have drawn enormous interest in recent times due to the biochemical relevance of its photoactivity. Photophysical properties of LH_2 in different solvents and solvent mixtures as well as its interaction with several biological molecules were reported in the literature (Vasilescu et al. 2003, Mitra, Das, and Mukherjee 1995, Guha et al. 1998, Banerjee, Mandal, and Mukherjee 2003, Nakamura and Nakamura 1998). Also, the fluorescence properties of LH_2 have been successfully applied in analytical and biochemical applications (Seitz and Rudolf 1975, Takayuki et al. 1986, Marilena Vasilescu et al. 2003).

Intermolecular hydrogen bonding and solubility of organic systems are known to play a crucial role in determining the biological activity as well as its application in forensic science (Jin et al. 2008, Committee 1992, Tsach, Wiesner, and Shor 2007, Miller and Bartick 2001, National Research 1992). In general, formation of intermolecular hydrogen bond between solute and solvent results in a decrease in Gibb's free energy and thus promotes mixing. Hydrogen bonding can occur in different modes, and critically depends on the structure of solute and solvent. The situation becomes more complicated when a solute molecule possesses multiple hydrogen bonding site and the solvent molecule can act both as a proton donor as well as proton acceptor.

Under this condition, the competition among different molecular species resulting from hydrogen bonding interaction between the solute and the solvent molecules becomes inevitable. LH_2 provides a unique example to study the hydrogen bonding effect because the molecule itself can exist in more than one prototropic structure (Figure 1) having multiple hydrogen bonding sites. The keto-amine (**I**) species can be converted into the enol-imine (**III**) form in single step or through the intermediate structures **IIa** and/or **IIb**, respectively. The inter-conversion among distinct species and associated spectroscopic properties will strongly depend on the relative abundance of individual component, as well as their hydrogen bonding mode with the solvent. Furthermore, the efficacy of hydrogen bond formation in the excited state may change due to charge redistribution after excitation. Also, relatively strong and unstructured fluorescence of LH_2 is an additional advantage to use it as a spectroscopic marker for quantitative estimation of the effect of different solvent parameters. The chemiluminescence and the fluorescence band of LH_2 in water appear in the same wavelength region (425~430 nm) (Vasilescu et al. 2003), so quantitative characterization of this band on different solvent parameters is indispensable.

Chemiluminescence behavior of LH_2 is being extensively used for quantitative detection of several proteins and cancer bio-markers by a variety of techniques (Zhang et al. 2009, Bi, Zhou, and Zhang 2009, Wu et al. 2008). Furthermore, LH_2 based chemiluminescent assay has been reported to detect the reactive oxygen species produced by human neutrophils (Shleev et al. 2008). Fluorescence is same as chemiluminescence, only exception being the excitation source (a light source is used to excite the electrons in the former instead of an appropriate chemical reaction in the later). Thus, one can monitor the variation in LH_2 fluorescence intensity to understand it's interaction with several biochemical environments. Also, the fluorescence properties of LH_2 have been successfully applied in different analytical and biochemical applications (Lasovský and Grambal 1986, Vasilescu et al. 2003, Seitz and Rudolf 1975, Takayuki et al. 1986). However, to be used as an efficient bio-marker assay, it is indispensable to study the fluorescence properties of LH_2 in presence of other biologically relevant medium like surfactants, cyclodextrin, proteins etc.

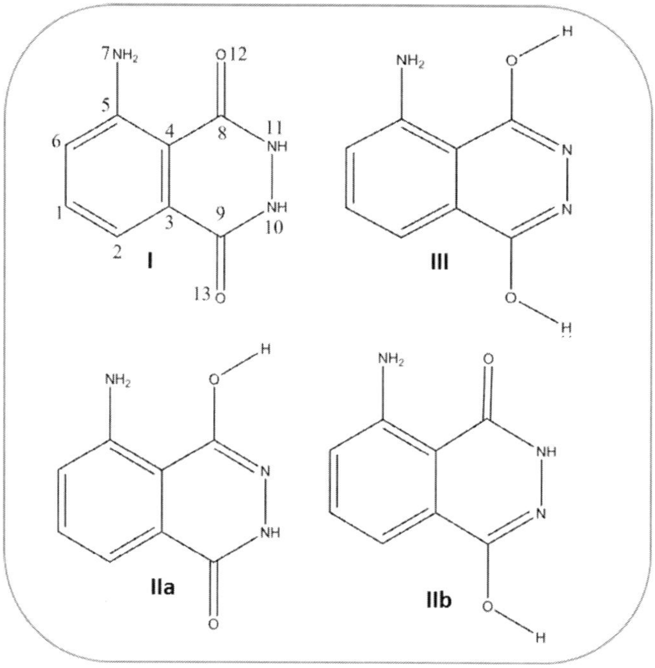

Figure 1. Different possible structures of Luminol (LH_2): Normal structure (**I**), mono-tautomeric structures (**IIa** and **IIb**) and di-tautomeric structure (**III**). The atom numbering used for DFT calculations are also shown.

It is already well-known that the chemiluminescence of LH_2 in H_2O_2 can be catalyzed by different metals nanoparticles (NPs) like silver, gold and platinum, metal oxide NPs like ZnO, ferric oxide, cerium oxide including gold/silver alloy and $CoFe_2O_4$ magnetic nanoparticles etc. (Guo et al. 2008, Chen et al. 2007, Zhang et al. 2005, Xu and Cui 2007, Li et al. 2009, Triantis et al. 2008, He et al. 2010). These NPs catalyze the decomposition of H_2O_2 to produce some reactive oxygen species, thus enhancing the CL from the reaction between LH_2 and H_2O_2. The CL of LH_2 - H_2O_2 can also catalyzed by different metal ions (like Co^{2+}, Cr^{3+}, and Cu^{2+}), organic compounds (like glutathione, L-cysteine, N-acetylcysteine), and albumin proteins (like bovine serum albumin) (Li et al. 2006, Du, Li, and Lu 2001, Hajime 1987).

Gold nanoparticles can also act as a catalyst for instantaneous generation of hydrogen peroxide (H_2O_2) from dissolved oxygen in luminol-hydrazine system and this sensitive method is used to detect hydrazine quantitatively in boiler feed water (Safavi, Absalan, and Bamdad 2008). Interestingly, the CL of LH_2-H_2O_2 system is not affected by NP size, but it depends on concentration of ions (Tsogas, Giokas, and Vlessidis 2014). Chemiluminescence resonance energy transfer (CRET) is a nonradiative (dipole-dipole) energy transfer from donor to an appropriate acceptor molecule. Normally, metal NPs are used as efficient energy acceptor due to the strong overlap of their absorption band with donor emission. Nanomaterial based CRET has application in immunoassays, like DNA and micro-RNA analysis, small molecules detection, cell imaging and used to design optical sensor as well as analytical methodology (Huang and Ren 2011, 2012).

Molecular properties of organic systems can be functionalized by physical adsorption on noble metal NPs towards desired applications including biomedical sciences (Niemeyer 2001, Rosi and Mirkin 2005), antibacterial systems (Hajipour et al. 2012), sensing elements (Upadhyayula 2012) as well as drug carriers (Agasti et al. 2010). Surface immobilized NP based energy transfer (ET) systems are very promising for a series of important biological applications including nucleic acid analysis, immunological bioassays, cancer cell detection, and drug delivery etc. Further, the application of metal NPs as acceptors in ET pair has significantly improved the overall efficiency of the process that enables to probe a larger conformational change of a macromolecule, which has been out of reach until now involving organic molecules as acceptors in the conventional ET systems. The primary reason for this enhanced sensitivity of ET is attributed to the electromagnetic interaction between the dipole of the donor fluorescence and the surface electrons of the metal nanostructure (Chatterjee, Lee, et al. 2011), that can take place even at sufficiently longer distance compared to traditional dipole-dipole coupling. Further, the geometrical restriction in a typical D – A pair seems irrelevant due to the spherical shape of the acceptor NPs. On the other hand, extensive literature reports are available on the modification of metal NP surface due to the adsorption of plasma proteins on it (Maleki, Moradi, and Tahmasebi 2017).

Immense interest is to probe the bovine serum albumin (BSA) adsorption on different NP surface because of several reasons like very high stability of the protein, availability at high purity and its solubility in water.

In the present contribution, the hydrogen bonding interaction of luminol (LH_2) and its interaction with several biological media are monitored with the help of steady state and time-resolved fluorescence as well as density functional theory (DFT) calculations. The layout of the present discussion has been organized as follows: a brief description on the experimental procedure is given next followed by the main results and discussion section. The first few sub-sections comprise in-depth understanding of solvent dependent photophysical behavior of LH_2 followed by the detailed information on modulatory fluorescence behavior in presence of a series of bio-mimicking environments like cyclodextrins, surfactants, proteins and DNA nucleotides etc. At last, the energy transfer between excited LH_2 as donor and spherical metal nanoparticle (MNP) as acceptor is discussed in presence of bovine serum albumin towards developing a more efficient biosensor.

2. Materials and Methods

Luminol (LH_2, 97%) was received from Sigma-Aldrich Chemical Pvt. Ltd. (product no. 123072) and used without any further purification (Moyon, Chandra, and Mitra 2010). Freshly prepared LH_2 solution of 5~10 μM concentration was used for different spectroscopic measurements. The details of synthesis and complete characterization of metallic gold and silver nanoparticles (Au and AgNP, respectively) with estimated size of 13±2 and 8±2 nm have been described elsewhere (Sonu et al. 2016). The concentration of synthesized nanoparticles is ~7.2 and ~2.6 nM for silver and gold, respectively (Sonu et al. 2016, Haiss et al. 2007).

Recent success of the DFT method (Benco et al., 2001; Chandra et al., 2004; Mavri et al., 2001; Parveen et al., 2009; Pan et al., 1997; Hao, 2006) in elucidating the complex phenomenon like hydrogen bonding prompts us to use it further in studying the energetic parameters of isolated LH_2 and its complex with water molecules. To elucidate the lowest energy structure, full geometry optimization was performed for all the possible conformers of LH_2, both in isolated (Figure 1) as well as in their mono- and di-hydrated complexes using the B3LYP method and 6-311++G (d,p) basis set as implemented in Gaussian09 program package. Frequency calculations were done in each stationary point to characterize the minimum energy equilibrium structure. Vertical transition energies up to the first ten singlet excited states and corresponding oscillator strengths for all the structures were estimated using time dependent DFT method (TD-DFT) at the same level of calculation.

Steady state absorption and fluorescence experiments were performed in Lambda25 (PerkinElmer) and FL-4500 (Hitachi) spectrophotometers, respectively. Since none of the components providing the bio-mimicking environment have any significant absorption at the excitation wavelength, any negligible attenuation of the incident light in the fluorescence (F) of LH_2 in presence of different concentration of the additives was corrected using the following equation (Valeur 2002):

$$F^{corr}(\lambda_E, \lambda_F) = F(\lambda_E, \lambda_F) \times \frac{A(\lambda_E)}{A_{tot}(\lambda_E)} \qquad (1)$$

Where, A represents the absorbance of the free LH_2 and A_{tot} is the total absorbance of the solution at the excitation wavelength (λ_E). Fluorescence quantum yields (ϕ_f) were calculated by comparing the total fluorescence intensity under the whole fluorescence spectral range with that of a standard (quinine bisulfate in 0.5 M H_2SO_4 solution, $\phi_f^s = 0.546$) using the following equation with adequate correction for solvent refractive index (n) (Mataga, 1970; Olson and Christ, 1996).

$$\phi_f^i = \phi_f^s . \frac{F^i}{F^s} . \frac{1-10^{-A^s}}{1-10^{-A^i}} . (\frac{n^i}{n^s})^2 \qquad (2)$$

where, A^i and A^s are the optical density of the sample and standard, respectively; and, n is the refractive index of solvent at 293 K. The relative experimental error of the measured quantum yield was estimated within ±5%.

Time-resolved fluorescence decay curves of LH_2 were collected by exciting the sample with 365 nm LED in a picomaster (PM3) time-correlated single photon counting (TCSPC) apparatus (PTI, USA) (Sonu et al. 2016). The experimentally obtained decay traces F(t) were expressed as a sum of exponentials (Eq. 3) (Bevington and Robinson 2003).

$$F(t) = \sum_i \alpha_i \exp\left(\frac{-t}{\tau_i}\right) \quad (3)$$

Where, α_i is the amplitude of the ith component associated with lifetime τ_i; such that $\Sigma\ \alpha_i = 1$. The average fluorophore lifetime (τ_{av}) was expressed as fractional contribution (f_i) of each decay time (Eq. 4) to the steady state intensity (Lakowicz 2006).

$$\tau_{av} = \sum_i f_i \tau_i = \frac{\sum_i \alpha_i \tau_i^2}{\sum_i \alpha_i \tau_i} \quad (4)$$

3. Results and Discussion

3.1. Steady State Spectral Properties and Fluorescence Solvatochromism

The fluorescence behavior of LH_2 was studied in a series of solvents with varying physico-chemical properties. Representative spectral profiles in aqueous medium is given in Figure 2a and Table 1 shows the steady state spectral parameters measured in different solvents. In general, LH_2 shows two absorption maxima with relatively large absorption coefficient; one at 280~320 nm region and the other is relatively broad and unstructured within 330~380 nm range. When excited at both the two absorption bands, LH_2

gives strongly intense, unstructured and broad emission spectra ranging from 375~550 nm. The excitation spectra show a broad peak near 350 nm corresponding to this emission. The origin of the low energy transition at 330~380 nm is assigned as $S_1(\pi) \leftarrow S_0(\pi)$; whereas, the high energy absorption at 280~320 nm corresponds to $S_2 \leftarrow S_0$ excitation. Theoretical calculations discussed in the later part of this chapter also confirm the nature and origin of this transition.

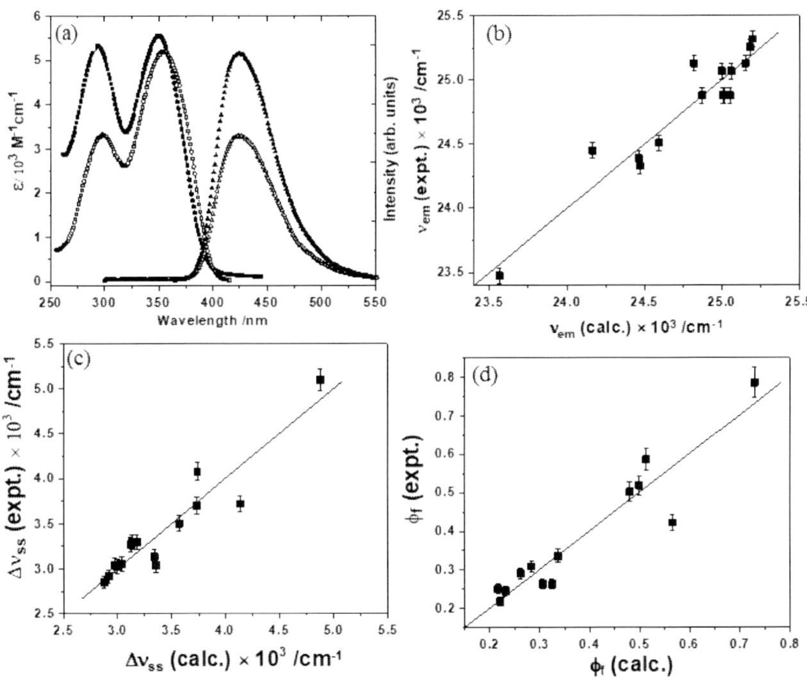

Figure 2. (a) Absorption (open square), fluorescence emission (open and filled triangle with λ_{exc} = 350 and 290 nm, respectively) and fluorescence excitation (filled square, λ_{exc} = 425 nm) spectra of luminol (LH_2) in water. (b, c and d) Correlation diagram of the luminol emission energy (ν_{em}), Stokes shift ($\Delta\nu_{ss}$), and fluorescence quantum yield (φ_f) with theoretically predicted values from Eq. 5.

Table 1. Steady state spectral properties of LH_2 in homogeneous solvents

Solvent[a]	$\Delta f(\varepsilon,n)$[b]	$E_T(30)$[c]	α[d]	β[d]	π^*[d]	$\nu_{abs}/10^4$, cm^{-1}	$\nu_{em}/10^4$, cm^{-1}	$\nu_{exc}/10^4$, cm^{-1}	$\Delta\nu_{ss}/10^3$, cm^{-1}	ϕ_f	$\delta_{em}/10^3$, cm^{-1}	$\delta_{exc}/10^3$, cm^{-1}
1. ACN	0.30	45.6	0.19	0.40	0.75	2.82	2.51	2.85	3.04	0.26	3.05	3.33
2. EAC	0.19	38.1	0.0	0.45	0.55	2.82	2.52	2.85	2.92	0.21	2.90	3.33
3. BEN	0.0	34.3	0.0	0.10	0.59	2.82	2.49	2.82	3.29	0.20	3.05	3.10
4. THF	0.21	37.4	0.0	0.55	0.58	2.82	2.53	2.84	2.85	0.24	2.90	3.26
5. DIO	0.03	36.0	0.0	0.37	0.55	2.82	2.51	2.85	3.04	0.25	2.85	3.71
6. TOL	0.02	33.9	0.0	0.11	0.54	2.82	2.49	2.84	3.29	0.21	3.16	4.01
7. DMSO	0.26	45.1	0.0	0.76	0.1	2.79	2.49	2.80	3.05	0.33	2.90	3.26
8. DMF	0.27	43.2	0.0	0.69	0.88	2.80	2.50	2.82	3.03	0.26	2.90	3.31
9. DCM	0.22	40.7	0.0	0.10	0.81	2.80	2.49	2.85	3.13	0.29	3.11	3.36
10. BuOH	0.26	49.7	0.84	0.84	0.47	2.84	2.43	2.80	4.08	0.52	3.00	3.59
11. MeOH	0.31	55.4	0.98	0.66	0.60	2.82	2.44	2.83	3.72	0.42	3.06	4.20
12. PrOH	0.27	50.7	0.84	0.90	0.52	2.81	2.43	2.80	3.70	0.58	3.12	3.77
13. PeOH	0.25	49.1	-	-	-	2.79	2.44	2.80	3.48	0.58	3.06	3.79
14. WAT	0.32	63.1	1.17	0.47	1.09	2.86	2.35	2.84	5.10	0.78	3.21	4.39
15. ACE	0.28	42.2	0.08	0.48	0.71	2.83	2.50	2.81	3.26	0.31	2.90	3.00
16. iPrOH	0.27	48.4	0.76	0.95	0.48	2.80	2.45	2.82	3.50	0.50	3.06	3.86

[a] Abbreviations used: ACN (acetonitrile), EAC (ethyl acetate), BEN (benzene), THF (tetrahydrofuran), DIO (1,4-dioxane), TOL (toluene), dimethyl sulfoxide (DMSO), DMF (dimethyl formamide), DCM (dichloromethane), BUT (1-butyl alcohol), MeOH (methyl alcohol, PrOH (1-propyl alcohol), PeOH (1-pentyl alcohol), WAT (water), ACE (acetone), iPrOH (isopropyl alcohol), ACN (acetonitrile), EAC (ethyl acetate). ν = absorption, emission and excitation energy; $\Delta\nu_{ss}$ = Stokes shift; ϕ_f = fluorescence quantum yield; δ = corresponding full width at half maxima (FWHM), [b]Polarity parameter ($\Delta f = \frac{\varepsilon - 1}{2\varepsilon + 1} - \frac{n^2 - 1}{2n^2 + 1}$) where, solvent dielectric constant and refractive indices are represented by ε and n, respectively; [c]Reichardt solvent parameter; [d]Kamlet-Taft solvent parameters.

The wavelength of fluorescence emission maximum (λ_{em}) of LH_2 shows appreciable shift in protic solvents along with almost two-fold increase in fluorescence quantum yield (ϕ_f) in comparison with that in aprotic solvents (Table 1). Interestingly, the spectral profiles both for the absorption and emission in aqueous and other protic media are relatively broader than in aprotic solvents. These results indicate that the interaction of LH_2 in water (or other protic solvents) are different and hydrogen bonding interaction seems to play a significant role in describing the photophysics of LH_2.

Solvent molecules can interact with organic solute molecules both through non-specific as well as specific interactions like hydrogen bond (HB) formation etc. Along with the polarity/polarizability induced non-specific modulation of spectral parameters, typically protic solvents can form HB both through HB donation (solvent acidity) or acceptance (solvent basicity). Nevertheless, the molecular propensity of solute solvent interaction can be reasonably estimated using solvatochromic studies. Solvent-dependent spectral properties are often interpreted in terms of solvent orientation polarizability (Δf) or a single empirical polarity parameter like $E_T(30)$. Unfortunately, none of the spectroscopic properties of LH_2 show any regular solvatochromism behavior with these parameters. This observation points to the existence of specific solute-solvent interactions; presumably both through HB donation acidity and HB acceptance basicity.

The contribution of several solvent parameters on the spectroscopic behavior of LH_2 are interpreted by means of the linear solvation energy relationship (LSER) through multi-parametric regression method (Eq. 5) developed by Kamlet and Taft (Kamlet, Abboud, and Taft 1981).

$$P = P^0 + s\pi^* + a\alpha + b\beta \quad (5)$$

Where, P is the value of the solvent dependent property to be modeled, P^0, s, a and b are the coefficients determined from the LSER analysis. The term π^* indicates the measure of solvent dipolarity/polarizability (Abboud, Kamlet, and Taft 1977); whereas, α and β is the scale of hydrogen bond donation acidity and acceptance basicity of the solvent, respectively (Miller and Bartick 2001). The parameters corresponding to the investigated solvents are taken from literature (Marcus 1993, Kamlet et al. 1983) and given in Table 1. The correlations of the spectroscopic data were performed by multiple linear regression analysis as implemented in ORIGIN 6.0 (Microcal Inc.) program package.

The *s*, *a* and *b* coefficients, obtained from LH_2 spectral data, are given in Table 2. Some representative correlation diagrams of the experimental values with those calculated from multiple regression analysis are shown in Figure 2 (b, c and d). Careful analysis of the results reveals several interesting feature for LH_2 solvatochromism: (i) in general, the contributions from *a* as well as *b* parameters are significant relative to the *s* parameter, thus confirming the importance of solvent hydrogen bonding in LH_2 spectroscopy; (ii) the excited state spectral properties like fluorescence maxima, Stokes-shift, quantum yield etc. are mostly controlled by solvent hydrogen bond acidity function (*a* parameter); whereas, both *a* and *b* contribute almost equally in the absorption property. This indicates an efficient charge localization in LH_2 upon excitation, as indeed confirmed by the DFT calculation (given below); (iii) the charge localization in the excited state is further confirmed by the negative values of both *a* and *s* (Velazquez et al. 1995); (iv) almost two fold increase in fluorescence quantum yield in protic medium (Table 1) is mainly due to the hydrogen bond acidity of the solvents with $a/s \sim 1.5$. However, there is insignificant effect on ϕ_f (~8%) for the solvents like DMSO which possesses higher hydrogen bonding acceptor (HBA) tendency. For example, the ϕ_f value of LH_2 in water is about 0.78 compared to that in 1, 4-dioxane (0.25) and DMSO (0.34). This prediction is in accordance with the experimental observation that LH_2 fluorescence yield decreases substantially in presence of water soluble proteins like bovine and human serum albumins (see below). Finally, (v) there is a large spectral shift in case of water and other hydroxylic solvents, as observed in Table 1. This is due to the larger negative contribution of 'a' parameter (Table 2). In conclusion, solvatochromic study shows that the fluorescence is predominantly controlled by the excited state of LH_2-solvent hydrogen bonded complex originated through the hydrogen bonding acidity behavior of protic solvents like water.

Table 2. Regression fit to solvatochromic parameters towards the steady state spectral properties of LH_2 and their relative values (in percentage)

Property (P)	P^0	s	a	b	P_s(%)	P_a(%)	P_b(%)
ν_{abs} /cm^{-1}	28183.14	70.06	260.90	-267.79	11.70	43.57	44.73
ν_{em} /cm^{-1}	25362.49	-652.76	-1448.0	-391.26	30.65	51.03	18.32
ν_{exc} /cm^{-1}	28492.90	25.60	26.92	-424.06	5.37	5.64	88.89
$\Delta\nu_{ss}$ /cm^{-1}	2820.65	722.80	1347.85	-657.84	26.49	49.38	24.13
ϕ_f	0.074	0.225	0.33	0.05	37.19	54.55	8.26
δ_{em} /cm^{-1}	3061.69	28.84	233.77	-277.26	5.34	43.30	51.36
δ_{exc} /cm^{-1}	3566.69	-30.11	851.47	-491.86	2.19	61.99	35.82

3.2. Identification of the Important Hydrated Species: DFT Calculation

LH_2 can have several types of tautomeric structures. The energy parameters of all these structures, obtained from DFT calculation, are shown in Table 3. Among all the conformers, the conformer **IIb** is found to be the most stable structure in the isolated, monohydrated (**IIb-S1** and **IIb-S2**) as well as in the dihydrated (**IIb-S3**) state (Figure 3). However, since the relative energy of **IIb-S3** is the lowest, LH_2 is most likely to exist in this form as compared with other hydrated structures. The relative energy of **I-S3** is only ~7.6 kJ mol^{-1} higher than **IIb-S3**. It suggests that the availability of structure **I-S3** is possible (while comparing with **IIb-S3**) in the ground state; and more so, in the excited state (see below). But the existence of other relatively higher energy conformers (**IIa** and **III**) can be ignored for discussing the spectroscopic behavior of LH_2 in water because they are very unlikely to be present in solution.

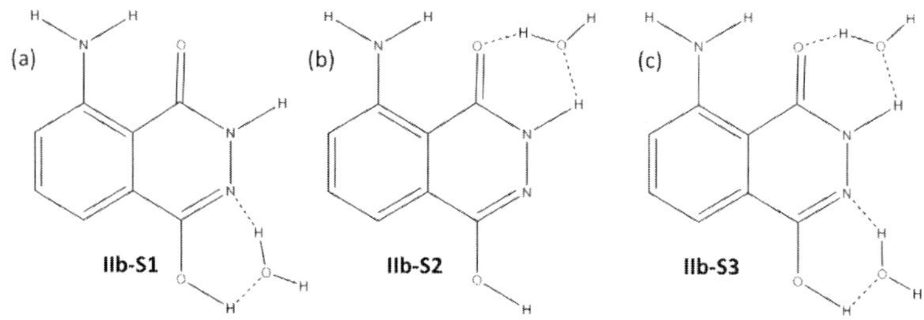

Figure 3. Schematic representation of mono-hydrated (**IIb-S1** and **IIb-S2**) and di-hydrated (**IIb-S3**) structure of the most stable conformation of luminol (**IIb**).

Table 3. Relative energy (kJ mol^{-1}) of fully optimized structures of different conformers of LH_2 in gas phase calculated at B3LYP/6-311++G(d,p) level

Structure[a]	I	IIa	IIb	III
Isolated	95.504	99.626	81.421	141.022
Monohydrated (S1)	52.952	58.062	41.553	90.538
Monohydrated (S2)	50.450	57.507	41.715	93.064
Dihydrated (S3)	7.635	14.544	0.0	44.182

[a] See Figures 1 and 3 for the structures of different conformers with varying degree of hydration.

Excited states for both the stable ground state structures of dihydrated LH_2 (**I-S3** and **IIb-S3)** were analyzed using TD-DFT procedure. The observed values of transition energies for different singlet excitations (within the experimentally observed absorption wavelength range 260~400 nm) and the corresponding oscillator strengths are given in Table 4. The nature of transition as well as its energy and oscillator strength are found to be comparable for both the structures. The calculated $S_1 \leftarrow S_0$ transition wavelength of ~335 nm agrees well with the experimentally observed value of 350 nm. However, the nature of the second lowest energy transition in the experiment cannot be assigned unambiguously as the energy separation of the next two calculated values (294 and 281 nm, respectively) are very close to each other with almost similar oscillator strengths. However, the associated orbitals, viz. HOMO-1, HOMO, LUMO and LUMO+1, which are involved in excitation with significant contribution within the given wavelength range (Table 4), are of pi type in nature (Figure 4). Comparison of the nature of HOMO and LUMO points to the fact that on excitation, the electron density is more localized on carbonyl oxygen and imino nitrogen atoms (O12 and N10, respectively in Figure 1), which in turn increases the basicity at these points significantly. This further confirms the importance of hydrogen bond donation ability (acidity) of the solvent for discussing the spectroscopic behavior of LH_2, more particularly in the excited state, as indeed predicted from the LSER analysis discussed above.

Table 4. Calculated singlet excited state transitions, associated energies and oscillator strength (*f*) of I-S3 and IIb-S3[a]

	I-S3			**IIb-S3**		
Singlet State	Transition	Energy /nm	*f*	Transition	Energy /nm	*f*
1	H→L	335	0.13	H→L	332	0.14
2	H-1→L	294	0.01	H→L+1	294	0.07
3	H-1→L+1 H→L+1	282	0.07	H-1→L H→L+1	281	0.05

[a] The frontier orbitals HOMO and LUMO are designated as H and L, respectively.

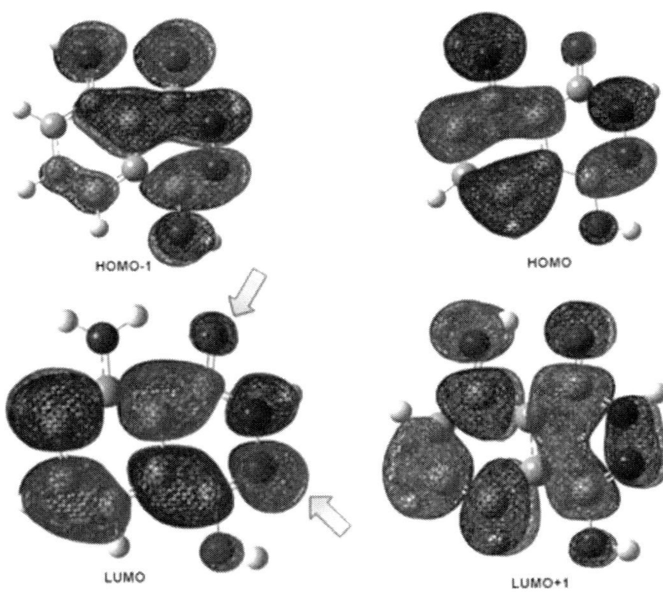

Figure 4. Frontier orbitals of **IIb** conformer of LH_2. The arrow mark in LUMO indicates the point of charge localization on electronic excitation.

The difference in the calculated energy between **I** and **IIb** in the ground state is ~14.1 kJ mol^{-1}. The potential energy surface (PES), constructed by intrinsic reaction coordinate (IRC) calculation from the transition state (TS), estimates a large activation barrier of ca. ~55 kJ mol^{-1} for the water assisted conversion from **I** to **IIb**. Therefore, the relative abundance of high energy structure **I-S3** is expected to be very less in the ground state, both from the kinetic and thermodynamic point of view. However, TD-DFT calculation results envisage a very small relative energy difference (~3.4 kJ mol^{-1}) between **I-S3** and **IIb-S3** in the first excited state. Hence, simple Boltzmann distribution predicts approximately 25% population of excited state to exist as **I-S3** in solution at room temperature. Therefore, the photochemistry of LH_2 can be accounted as an average property coming from both the structures present in the excited state. The transition energy, nature of excitation as well as the corresponding oscillator strength are alike for both the structures (Table 4). Naturally, it is expected that the above two conformers will show similar spectroscopic behavior, predominantly in non-interacting solvents. However, as these conformers are different in their mode and extent of hydrogen bonding, there is possibility to form different hydrogen bonded clusters in protic solvents with minor difference in energy. The broad emission spectra of LH_2 in protic solvents may be due to the ensemble averaged spectral properties of all these microstructures.

Picosecond time-resolved fluorescence decay measurement of LH_2 emission in aqueous buffer solution indicates a bi-exponential decay function as confirmed by different statistical parameters and visual inspection of the fitting data. It is observed that about 25% of the total excited fluorophores decay at about 2.3 ns; whereas, the remaining population has an average fluorescence decay time of ~7.5 ns. The measured TCSPC data is in good agreement with the decay times of LH_2 reported in the literature (Mitra, Das, and Mukherjee 1995). DFT calculation result discussed above indicates that only structures **I** and **II** (Figure 1) are responsible for the fluorescence behavior of LH_2 in solution (Moyon, Chandra, and Mitra 2010) with structure **II** is more stable by ~14.1 kJ mol^{-1} in energy compared to structure **I**. Therefore, the short and long decay time component of LH_2 fluorescence can be assigned to structures **I** and **II**, respectively.

3.3. Modulation of Luminol Fluorescence in Micellar Media

Since homogeneous mixture of water and 1,4–dioxane closely mimics the micellar environment, it is worth discussing the spectral behavior of LH_2 in these mixtures first. With increase in the volume fraction of water, the fluorescence spectral position exhibits a red shift of about 30 nm. The fluorescence quantum yield (ϕ_f) also increases about three times in pure aqueous medium while comparing with that in 1,4-dioxane. Figure 5 shows representative emission spectral profiles along with the corresponding data. This notable change in the fluorescence properties of LH_2 in water is believed to be due to the stabilization of excited state(s) by the formation of specific hydrogen bond between LH_2 and water.

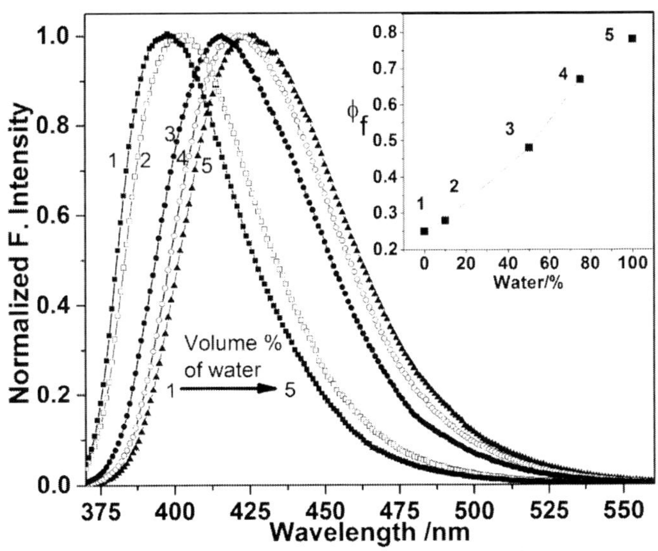

Figure 5. Normalized fluorescence emission spectra of LH_2 in varying water/1,4-dioxane mixtures. The volume percent of water = 0 (1), 10 (2), 50 (3), 75 (4) and 100 (5). Inset shows the quantum yield (ϕ_f) vs percentage of water contained.

The absorption maximum of LH_2 does not undergo any noticeable change on addition of surfactants to the aqueous medium. In contrast, the intensity of the fluorescence is greatly altered with the increasing amount of surfactant in solution. Continuous addition of all the tested surfactants (CTAB, TX-100 and SDS) causes an initial increase within very small surfactant concentration, followed by a regular decrease in LH_2 fluorescence intensity. The representative spectral profiles are shown in Figure 6(a, b, c) along with the variation in the intensity pattern in each case. The increase in the fluorescence intensity at very low surfactant concentration can be attributed due to the increase in solubility of organic fluorophore; however, the fluorescence intensity decreases continuously after a certain point where the concentration of the surfactant is very close to the critical micelle concentration (cmc) of the individual surfactant system. Due to the formation of the micellar structure, the fluorophore moves into the hydrophobic micellar pseudo-phase from bulk aqueous medium resulting a decrease in average steady state fluorescence intensity, which is consistent with the observations made above for homogeneous solvents like water buffer and 1,4-dioxane. Interestingly, from the inset diagram Figures 6a and 6c, it is evident that the rate of decreases in the fluorescence intensity after the micelle formation is relatively higher in case of cationic micellar system CTAB, as compared with anionic SDS. However, it is rather difficult to compare quantitatively the TX-100 data given in inset of Figure 6b, since the range of the concentration used are at least an order of magnitude lower than the other two. Apparently, the peak position of fluorescence emission is not too sensitive towards the micellar medium. But careful analyses of the data presented in Table 5 and Figure 6 confirm that fluorescence spectral blue shift is slightly greater in CTAB (~3 nm) while comparing with other surfactant systems. In analogy with the results discussed above for water/1,4-dioxane solvent mixture, a blue shift of the fluorescence maximum suggests that the polarity of the micellar environments is less than the polarity of the bulk water. However, the data for fluorescence maxima of LH_2 in fully micellized condition (Table 5) indicates that the probe is not fully incorporated into the core; rather, it stays in the interfacial region of the micelle. The decline in LH_2 fluorescence intensity in micellar medium can be explained on the basis of the passage of the relatively non-polar fluorophore (structure **II** with dipole moment μ = 1.92 D in comparison with **I** having μ = 2.12 D) towards more hydrophobic interfacial region. However, due to greater charge localization in the excited state, LH_2 interacts more strongly with cationic micelle like CTAB. The existence of the dipole-dipole type of interaction in addition to the hydrophobic force causes much stronger interaction with CTAB in comparison with other surfactant systems as evidenced by (a) rapid fall in fluorescence intensity, and (b) detectable blue shift in fluorescence spectral position. This is further confirmed by studying the time resolved fluorescence as well as thermodynamics of ligand binding in BSA, discussed below. It is to be noted here that while discussing the principle driving force and the stability of LH_2-micelle complex in different surfactant systems, neither the dimension of the micelle nor the aggregation number was taken into consideration. Also, a rigorous description of the binding forces responsible for binding the chromophore towards the micellar sub-domain should involve the size of the respective micelles and its change with increasing surfactant concentration as well as the spatial distribution of the fluorophore in the interfacial region. Nevertheless, a qualitative description based on the averaged parameters of the probe and micellar system is quite reasonable to explain the tendency of the hydrophobic environments in quenching of steady state LH_2 fluorescence intensity and the time-dependent fluorescence decay behavior (discussed below).

The fluorescence decay behavior of LH_2 was also monitored in the presence of added surfactants. Again, the observed experimental data could be well reproduced with two-exponential decay functions and Table 5 displays all the relevant data for different surfactants. From the table, it is quite clear that the fluorescence decay time comprising two components does not change in presence of all the surfactants. When the concentration of the surfactants is above cmc, the relative contribution of the first decay component is found to increase. However, this increase is almost two and half times in case of CTAB as compared to the other surfactants. This observation further supports the idea of relatively stronger interaction in CTAB micelle as discussed before.

Table 5. Fluorescence spectral behavior of LH_2 (~ 5 μM) in homogeneous buffer solution and in different heterogeneous medium[a]

Medium		λ_{max}^{em} (nm)	ϕ_f	Decay parameters		
				τ_1 /ps (α_1)	τ_2 /ns (α_3)	τ_3 /ns (α_3)
Buffer (pH=7.4)		426	0.78	-	2.4 (0.24)	9.8 (0.76)
CTAB	0.5 mM	426	0.84	-	2.5 (0.24)	9.9 (0.76)
	12.5 mM	423	0.61	-	2.4 (0.61)	9.5 (0.39)
TX-100	0.15mM	426	0.89	-	2.3 (0.25)	9.7 (0.75)
	0.6 mM	426	0.77	-	2.4 (0.36)	9.6 (0.64)
SDS	2.4 mM	426	0.81	-	2.4 (0.24)	9.9 (0.76)
	20 mM	426	0.59	-	2.5 (0.34)	9.5 (0.66)
BSA	6 μM	423	0.30	<100 (0.65)	1.8 (0.12)	8.9 (0.23)
	30 μM	420	0.11	<100 (0.60)	1.0 (0.27)	8.8 (0.13)
HSA	6 μM	426	0.62	<100 (0.71)	2.0 (0.08)	9.0 (0.21)
	30 μM	426	0.53	<100 (0.73)	1.6 (0.09)	9.0 (0.18)

[a]The pre-exponential factor and the decay time is represented by α and τ, respectively; the relative error in measuring τ values is within ±0.1 ns.

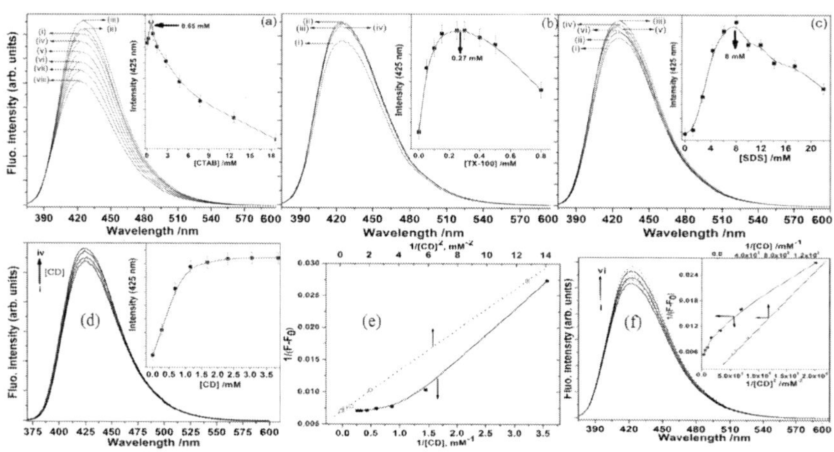

Figure 6. Fluorescence emission spectra of LH_2 (λ_{exc} = 360 nm) with increasing concentration of CTAB (a), TX-100 (b), SDS (c) and cyclodextrin, CD (d). Inset is the variation in fluorescence intensity monitored at 425 nm in each case. Double reciprocal plot for the interaction with CD in case of LH_2 (e) and LH_2-CTAB (f) system.

3.4. Fluorescence Behavior in β-Cyclodextrin (CD) and Mixed CD-Surfactant System

Cyclodextrin (CD) possesses the unique property to encapsulate organic compounds inside its hydrophobic central cavity. Therefore, these are considered as potential molecular vehicles for extremely efficient drug delivery (Szejtli 1985). Furthermore, the interior cavities of CD have less polarity and restricted geometry which gives an opportunity to study different photophysical properties in modified environmental conditions. The inclusion of the organic probe is primarily controlled by the size fitting of the host cavity toward the guest (Matsushita et al. 2004). The fluorescence intensity of LH_2 is found to increase slightly on addition of CD till it reaches a plateau following a spectral blue shift of about 5 nm (Figure 6d). As seen in the previous section about the LH_2 fluorescence quenching in micellar medium, it is rather surprising to observe an increase in the fluorescence intensity on binding with the hydrophobic CD cavity. In fact, slight increase in molecular fluorescence of LH_2 derivatives in presence of CD was reported as early as in mid-eighties; however, no explanation is given for that (Karatani 1986). Maeztu et al. also reported the enhancement of chemiluminescence intensity of LH_2 and its derivatives at alkaline pH in presence of natural cyclodextrins (Maeztu, Tardajos, and González-Gaitano 2010, Maeztu, González-Gaitano, and Tardajos 2010). Apart from the size requirement of the guest molecule for entire or partial inclusion inside the CD cavity, additional hydrophobic forces are also important in determining the geometry of the complex (Ogino 1981, Wylie and Macartney 1992). Nevertheless, the increase in fluorescence intensity of LH_2 in presence of CD may be due to the restricted non-radiative motion, and thereby reducing the vibrational deactivation of the caged analyte.

The apparent binding constant and stoichiometric ratio of the inclusion complex can be determined from the modified Benesi-Hilderbrand (BH) equation using the fluorescence data. The equations for the formation of 1:1 and 1:2 guest-host complex can be written in the following forms (Connors 1987).

$$\frac{1}{F-F_0}=\frac{1}{F_\infty-F_0}+\frac{1}{K_{a1}(F_\infty-F_0)}\times\frac{1}{[CD]} \tag{6a}$$

$$\frac{1}{F-F_0}=\frac{1}{F_\infty-F_0}+\frac{1}{K_{a2}(F_\infty-F_0)}\times\frac{1}{[CD]^2} \tag{6b}$$

Where, F_0 and F are LH_2 fluorescence intensities in absence and presence of CD, respectively. F_α indicates the fluorophore intensity at the plateau and K_a indicates the association constant in the respective cases. The double reciprocal plot for LH_2-CD system is shown in Figure 6d. The linearity for 1:2 plot explains that LH_2 is bound with two CD molecules and the binding constant is calculated to be 4.2×10^6 M^{-2} from the slope and intercept of the simulated data. Interestingly, there is also an increase in fluorescence intensity upon successive addition of CD in LH_2 solution containing a constant concentration of CTAB (12.5 mM). However, BH plot in this case indicates a complex with 1:1 stoichiometry with an apparent binding constant of 101.4 M^{-1}. This difference in stoichiometry may be due to the nature and intensity of the feeble forces responsible for the formation of the complex changes in presence of an ionic micellar medium like CTAB.

3.5. Fluorescence Modulation in Presence of Serum Albumin Proteins

The emission spectra of LH_2 in presence of various concentration of human and bovine serum albumin (HSA and BSA, respectively) are shown in Figure 7. With regular increase in the concentration of proteins, the fluorescence intensity of LH_2 are found to be decreased. However, there is a significant difference in the fluorescence maxima of LH_2 in the two cases. While a blue shift of ~10nm is observed in BSA, no significant shift in the emission maxima of LH_2 is noted in case of HSA. Further, the decrease in fluorescence intensity shows a clear sigmoidal shape in BSA. This is in sharp contrast to the observation in HSA, where the nature of fluorescence intensity decrease is non-sigmoidal. In general, the decrease in LH_2 fluorescence intensity can be assumed again as due to the passage of probe molecule from aqueous bulk phase and binding towards a more hydrophobic region of the ligand binding domain on the protein surface. Binding of LH_2 in albumin is further reinforced by ligand replacement process in presence of bilirubin (BIL). On successive addition of BIL to an LH_2/BSA solution, the LH_2 fluorescence intensity is recovered. BIL is known to possess strong affinity towards the albumins and binds with apparent binding constant varying within the range of approximately $10^6 \sim 10^8$ M^{-1} (Petersen et al. 2000). This value is higher by almost three orders in magnitude compared to that in case of LH_2 with BSA (see below). With the preferential incorporation of BIL into the ligand binding site, the bound LH_2 is expelled in the more aqueous phase; consequently, the fluorescence intensity increases.

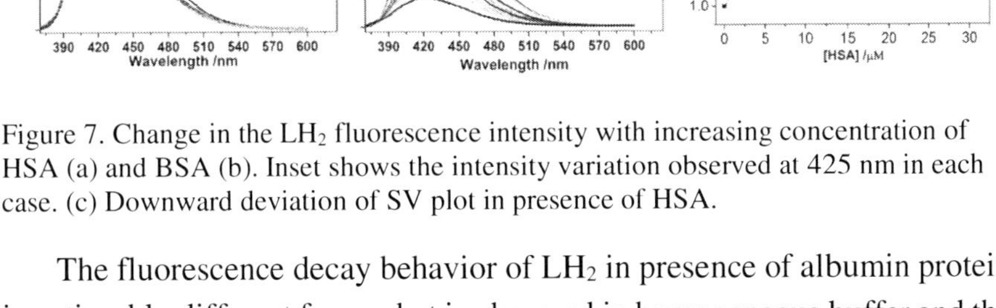

Figure 7. Change in the LH_2 fluorescence intensity with increasing concentration of HSA (a) and BSA (b). Inset shows the intensity variation observed at 425 nm in each case. (c) Downward deviation of SV plot in presence of HSA.

The fluorescence decay behavior of LH_2 in presence of albumin protein is noticeably different from what is observed in homogeneous buffer and the micellar medium (Table 5). Although the decay time corresponding to species **I** and **II** is still found to be ~2 ns and ~9 ns respectively, a major portion (ca. 65%) of the excited fluorophore decays with very short instrument limited lifetime (<100 ps). The presence of the first decay component in presence of BSA is obvious from Figure 8. Furthermore, a comparison of the statistical parameters, like the value of reduced chi-square and visual inspection of the weighted residual, reveals that two-exponential fitting of the experimental data is insufficient and a minimum of three-exponential decay function is required. In contrast, no additional fast decay was observed for LH_2 in presence of other heterogeneous micellar medium as discussed above (Table 5). Similar observation is also made in presence of HSA and can be explained on the basis of the formation of closely spaced fluorophore-quencher pair; where, the fluorescence is quenched rapidly in presence of the quencher and appears to be dark under this experimental condition (Moyon and Mitra 2010). This rapid transient quenching is associated with the additional fast decay time which is followed by the normal (slower) diffusion of the fluorophore and the quencher.

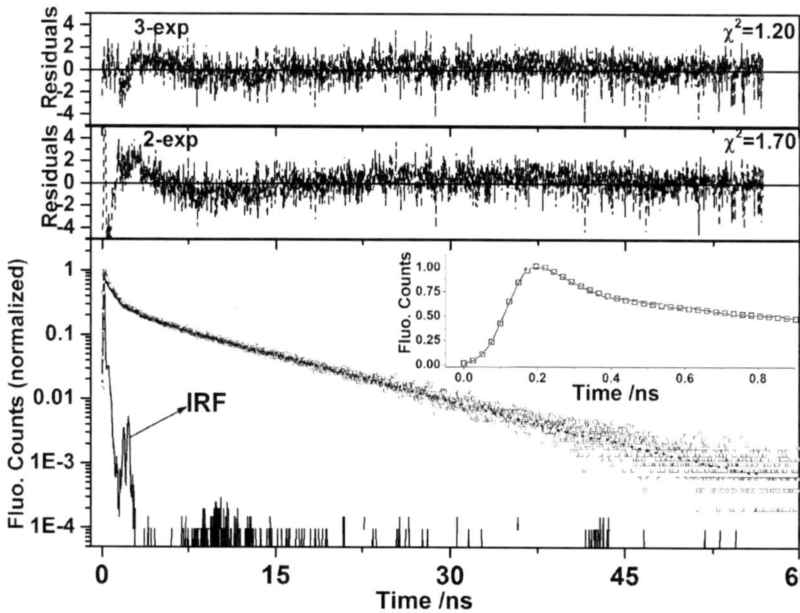

Figure 8. Fluorescence decay spectra (open circles) and simulated data (solid line) of LH_2 in presence of 30 μM BSA with three exponential decay function (λ_{exc} = 375 nm). IRF is the instrument response function. Inset shows the same data in shorter time scale. Upper panels show the distribution of weighted residuals and χ^2 values obtained from the fitting of the experimental data using three exponential (3-exp) and two exponential (2-exp) decay models, respectively.

The fluorescence quenching data was analyzed by Stern-Volmer (SV) model (Eq. 7) for both static and dynamic quenching (Valeur 2002).

$$\frac{F_0}{F} = 1 + K_{SV}[Q] = 1 + k_q \tau_0 [Q] \tag{7a}$$

$$\frac{\tau_0}{\tau} = 1\ (static);\ \frac{\tau_0}{\tau} = \frac{F_0}{F}\ (dynamic) \tag{7b}$$

The parameters F_0 (τ_0) and F (τ) indicate the fluorescence intensity (lifetime) in the absence and presence of certain quencher concentration [Q], respectively. K_{SV} is SV quenching constant and κ_q is diffusion controlled bimolecular quenching rate constant.

A plot of F_0/F of LH_2 versus [BSA] exhibits a good linearity (R = 0.994) and affords K_{SV} value $2.42 \pm 0.01 \times 10^5$ M^{-1} at 298 K. The amplitude weighted average lifetime in absence of quencher, $\langle\tau\rangle_0$, can be calculated from the bi-exponential decay data of LH_2 in homogeneous buffer (Table 5) and found to be about 8.0 ns. From this data, the bi-molecular quenching rate constant, κ_q, is calculated as 3×10^{13} M^{-1} s^{-1}. This value is almost three orders of magnitude higher than the maximum diffusion limited quenching rate constant, which is known to be of the order of 2×10^{10} M^{-1} s^{-1}(Ware 1962). Therefore, the quenching process is presumed mainly to be controlled by static quenching mechanism, rather than dynamically controlled process. For a static quenching process, K_{SV} can be regarded as the association constant (K_a) for the formation of the fluorophore-quencher complex in the ground state (Lakowicz 2006). The idea of static quenching is further supported by doing temperature variation experiment. The calculated association constant values from the linear SV plots (Figure 9) are collected in Table 6. It is seen that the magnitude of this quantity decreases with rise in temperature, which further supports the formation of protein-LH_2 complex.

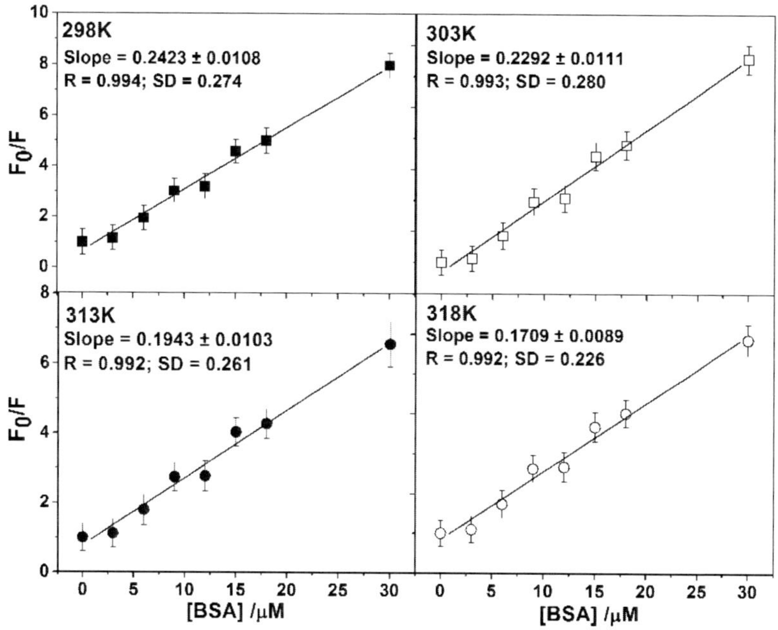

Figure 9. Stern-Volmer (SV) plots of LH_2 fluorescence quenching in presence of BSA at different temperatures. Linear fitting parameters and slope are shown in each case.

Table 6. Thermodynamic parameters corresponding to binding of LH_2 with HSA and BSA[a]

Temperature /K	Binding constant /M^{-1}	ΔH /kJ mol^{-1}	ΔS /J K^{-1} mol^{-1}	ΔG /kJ mol^{-1}
For HSA				
298	50.12	-31.70	-74.90	-9.38
303	28.86			-9.00
308	32.48			-8.63
313	22.17			-8.26
318	20.99			-7.88
For BSA / 10^5				
298	2.42±0.10	-13.52	57.99	-30.81
303	2.29±0.11			-31.10
308	2.27±0.16			-31.39
313	1.94±0.10			-31.68
318	1.70±0.08			-31.97

[a] The binding constant values were calculated from the slope of equation (7) and the other thermodynamic parameters were obtained using equations (10) and (11).

On the other hand, the SV plot for the LH_2 fluorescence quenching in HSA (shown in Figure 7) shows a downward curvature. This kind of deviation from linear SV plot is a typical characteristic feature of fluorescence quenching when two types of fluorophore populations are present; one of which being inaccessible to the quencher (Lakowicz 2006). A modified form of SV equation, given in Eq. (8), can be used to analyze this type of fluorescence quenching data assuming the total fluorescence to be the contribution from both the accessible and inaccessible fluorophore.

$$\frac{F_0}{\Delta F} = \frac{1}{f_a K_a [Q]} + \frac{1}{f_a} \tag{8}$$

Where, ΔF (= $F_0 - F$) is the difference of fluorescence intensity in absence and presence of quencher concentration [Q]; K_a is SV quenching constant of the accessible fraction (f_a).

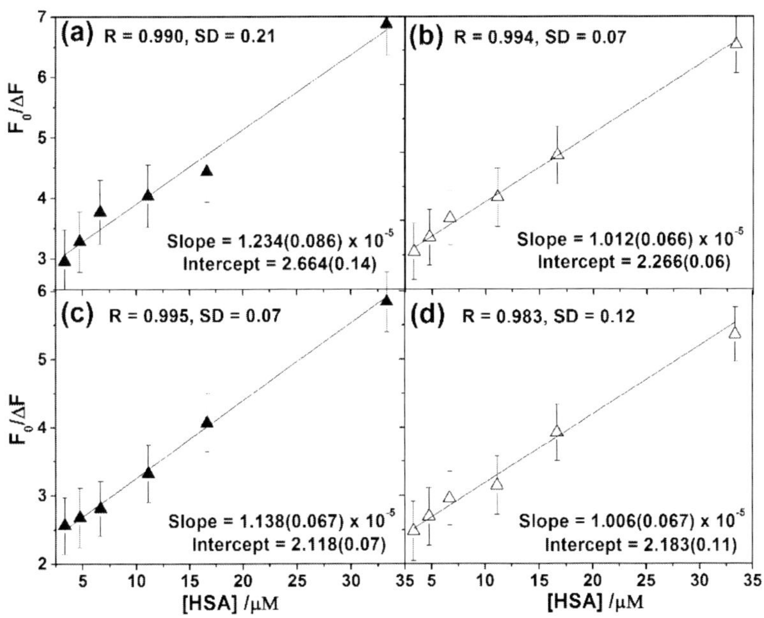

Figure 10. Modified Stern-Volmer plot for LH_2 fluorescence quenching by HSA at 298 K (a), 303 K (b), 308 K (c) and 313 K (d).

The values of K_a and the fraction of fluorophore accessible to the quencher (f_a) are obtained from the slope and intercept of the plot of $F_0/\Delta F$ vs $[Q]^{-1}$. Some of the representative linear variations are shown in Figure 10 at different temperatures in case of LH_2/HSA system. Apparently, the magnitude of the SV quenching constant ($2.0 \pm 0.2 \times 10^5$ M^{-1}) and fraction (40~47%) of the accessible fluorophore for quenching does not change significantly within the experimental temperature range.

The idea of fractional accessibility of LH_2 towards binding with HSA and consequent fluorescence quenching is further supported from the steady state and time resolved fluorescence study in presence of different micellar medium. The fluorescence intensity of LH_2 is found to decrease in presence of different surfactants like SDS, CTAB and TX-100, when the concentration of surfactants in solution is above critical micelle concentration (cmc). Moreover, time-resolved fluorescence decay analysis reveals that although the decay time on the first component (**I**) remains almost same (~2 ns), the long component (**II**) decays relatively faster in presence of micellar environment. Organic fluorophores are known to penetrate into the micellar core or can reside in the micelle-water interface when the surfactant concentration in solution is above cmc and therefore, experiences a relatively non-polar environment surrounding it (Zana 1986). In case of LH_2, the passage of component **II** in the micellar environment results in the reduction of both the intensity as well as fluorescence decay time of this species. However, component **I** remains in the homogeneous (aqueous) phase of the micellar solution and the decay time remains unaltered. Although the calculated dipole moment of species **I** and **II** (2.12 and 1.92 D, respectively) is not too different to conclude about the preferential encapsulation of species **II** in the relatively hydrophobic micellar core with certainty; however, it gives a qualitative indication towards this prediction and explains the observed experimental results.

HSA is known to possess a hydrophobic portion on its surface (sub-domain IIA), generally interior to the protein. Many hydrophobic ligands like fatty acids, bilirubin, and several indole derivatives like tryptophan bind at this region (Foster et al. 1965, Peters 1985). Out of the two conformers of luminol (**I** and **II**) present in solution, binding with serum albumin occurs through the more accessible hydrophobic species **II** and this in turn quenches the overall fluorescence intensity. Ligand replacement in presence of bilirubin (BIL) further reinforces the indisputable importance of sub-domain IIA in binding with LH_2. Further, gradual addition of BIL to an LH_2/HSA solution increases the fluorescence intensity of LH_2 continuously. As discussed before, BIL has a very strong affinity towards sub-domain IIA of HSA with apparent binding constant ranges within ca. $10^6 \sim 10^8$ dm^3 mol^{-1} (Petersen et al. 2000). This value is almost four times greater magnitude wise in comparison to that of LH_2 (see below). The fluorescence intensity increases because preferential binding of BIL into the ligand binding site which causes LH_2 to be expelled in the more aqueous phase. Also, the values of K_a in the LH_2 – BSA binding is about four orders of magnitude more than the corresponding value for LH_2-HSA binding. This indicates that the interaction of LH_2 with BSA is many times stronger than in the case of HSA.

The following equation depicts the equilibrium between the free and bound ligands in the cases where small molecules bind to a set of equivalent sites of a macromolecule (Yang et al. 2005):

$$\log\left[\frac{\Delta F}{F}\right] = \log K + n.\log[Q] \quad (9)$$

Where, K denotes the apparent binding constant to a site and *n* is indicative of the number of binding sites per macromolecule. ΔF ($=F_0 - F$) is the difference between the values of fluorescence intensities in the absence and presence of quencher (denoted by *Q*), respectively. Linear dependence of $log(\Delta F/F)$ against $log[Q]$ is observed when the experiment is carried out at varying temperatures for LH_2–HSA interaction. The correlation coefficient (R) obtained from data fitting is 0.98±0.01; while, the standard deviation (SD) values are found to be within 0.05. It further confirms that the assumptions made in deriving Eq. (9) hold true under the present experimental condition.

The thermodynamic parameters of LH_2-protein interaction have been calculated using the values of binding constant. In cases where the enthalpy (ΔH) does not undergo a significant change within the studied temperature range, van't Hoff relation (Eq. 10) can be used for evaluating the change in enthalpy and entropy for LH_2 binding to HSA (Silbey and Alberty, 2002). The corresponding Gibb's free energy parameter can be evaluated using the relationship given in Eq. (11).

$$logK = -\frac{\Delta H}{2.303RT} + \frac{\Delta S}{2.303R} \quad (10)$$

$$\Delta G = \Delta H - T\Delta S \quad (11)$$

All the thermodynamic parameters are listed in Table 6. The negative values of ΔG emphasize that the binding process is spontaneous within the entire temperature range. The magnitude for the changes in both enthalpy and entropy for LH_2 binding to HSA are seen to bear negative values. The interaction between a drug and bio-macromolecule may be due to several factors like electrostatic forces, van der Waals interaction, multiple hydrogen bond formation, steric contact within the cavity and hydrophobic forces etc. Ross and Subramanian (Ross and Subramanian 1981) have studied a series of protein binding reactions in detail with wide variety of ligands and proposed some general hypothesis from the data. According to them, three possible reasons may be associated with negative values of ΔH and ΔS. These are (i) non-bonded (van der Waals) interactions, (ii) formation of hydrogen bond in media with low dielectric constant and, (iii) protonation occurring during the association process. The probability of protonation during the LH_2-HSA association at the experimental pH is very low. Moreover, it has already been stated that LH_2 possesses several potential hydrogen bonding sites, and the hydrogen bonded conformers are much more stable thermodynamically (~80 kJ mol^{-1} for species **II**) than the corresponding non-hydrogen bonded conformers. Therefore, hydrogen bond formation even in the interior of the protein binding domain is a possibility that cannot be ruled out completely. Overall, it is an impossible task to account for all the thermodynamic parameters of LH_2-HSA interaction based only on a single intermolecular force model. A combination of several forces such as van der Waals interaction, possible hydrogen formation etc. may play important roles in the binding mechanism. Meanwhile, Table 6 shows that the major contribution in ΔG appears from ΔH rather than ΔS term; hence, the binding process is believed to be enthalpy driven.

The negative values of ΔG ascertain the binding of LH_2 with BSA being spontaneous within the whole temperature range. Further, the magnitude of ΔG for LH_2-BSA complex formation is found to be about three times than that of the interaction with HSA, justifying the assumption of stronger binding in the former. The high negative value of ΔH indicates that the binding is an enthalpy driven process. However, the positive entropy change asserts the fact that structural distortion also plays a vital role in the binding process. The results indicate that although BSA and HSA are structurally similar, the mechanism of binding and the corresponding thermodynamics of LH_2 interaction with these carrier proteins are distinctly different (see below).

LH_2 fluorescence quenching in presence of HSA exhibits a downward curvature in the SV plot which is explained because of fractional accessibility of the fluorophore within the protein's cavity. However, in case of BSA no such curvature is observed. This difference can be attributed to the location of the binding sites of the albumins, which are different in case of both the proteins. BSA, despite being structurally homologous to HSA, contains an extra binding pocket (sub-domain IA) which is comparatively more exposed to the aqueous medium in addition to a more hydrophobic binding region in sub-domain IIA for both the proteins (Carter and He 1990, He and Carter 1992, Dockal, Carter, and Ruker 1999, Carter and Ho 1994, Sudlow, Birkett, and Wade 1975, 1976, Zhong, Douhal, and Zewail 2000, Chuang and Otagiri 2006). The comparatively non-polar conformer of LH_2 binds in the hydrophobic binding region within sub-domain IIA of HSA (Moyon and Mitra 2010). However, the exposed binding region in sub-domain IA in case of BSA is rendered available for the relatively more polar fraction of LH_2, in addition to the hydrophobic binding in sub-domain IIA for non-polar fraction as seen in the case of binding with HSA. This is the reason that the overall fluorescence quenching obeys a simple SV relation in presence of BSA.

Both the binding sites in BSA are assumed to be cooperative in nature which results in a sigmoidal decrease of fluorescence intensity with increasing quencher concentration. Fluorescence spectral blue shift of ca. 6 nm in BSA, when compared to the bulk aqueous medium (Figure 7), further confirms the strong nature of LH_2 binding. In contrast, HSA shows no apparent spectral shift. Also, the comparative quenching of LH_2 emission intensity is seen to be almost five times higher in BSA compared to HSA in similar quencher concentration.

Modified Hill relationship (Eq. 12) is often used in quantitative description of cooperative binding process and analysis of the LH_2 fluorescence quenching in presence of BSA (Chatterjee, Mukherjee, et al. 2011, Pratap, Mikhaylyants, and Olden-Stahl 2009).

$$\Delta F = \Delta F_{max} \times \frac{[S]^{n_H}}{K_{0.5}^{n_H} + [S]^{n_H}} \tag{12}$$

Where, ΔF ($= F_0 – F$) denotes the change in fluorescence intensity in presence of substrate concentration [S] and ΔF_{max} represents the maximum change in fluorescence intensity. $K_{0.5}$ is the concentration of substrate that gives half maximal change in fluorescence intensity and n_H is the Hill coefficient. n_H usually represented by a whole number and when greater than one, the situation indicates positive cooperativity; on the other hand, n_H being equal to one represents a situation which has no cooperativity.

Figure 11 (right panel) gives the simulation of LH_2-BSA data using Eq. (12) at different temperatures. The experimental data points correlate nicely and the statistical parameters are acceptable with correlation coefficient greater than 0.99 in all the cases. It is interesting to note that the magnitude of n_H is very close to two, which indicates positive cooperativity between the two binding sites of BSA. However, similar analysis in case of HSA (Figure 11, left panel) estimates the value of n_H to be close to one, indicating the absence of any cooperative mechanism. A massive structural change in a cooperative manner occurs during the initial binding of LH_2 in the more exposed ligand binding site of BSA. This change in structure happens because the hydrophobic binding site in sub-domain IIA also gets easily accessible for LH_2 binding. The overall positive change in ΔS value manifests this reorganization in the protein structure on ligand binding. Further, the cooperativity is exemplified in almost four orders of magnitude higher value of binding constant and about three times greater negative free energy change in BSA binding when compared with HSA.

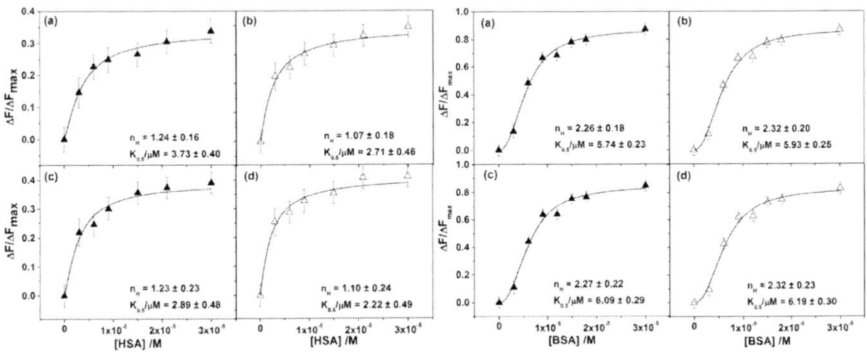

Figure 11. Hill plot for both LH_2-HSA (left panel) and LH_2-BSA (right panel) systems at 298 K (a), 303 K (b), 313 K(c) and 318 K (d). Solid line is the simulated data from the experimental using equation (12). The related parameters are also shown after the fitting.

3.6. Interaction of LH_2 with DNA Nucleotide Bases

To understand the interaction of luminol with DNA, fluorescence experiments of LH_2 were performed with two DNA nucleotide purine bases like adenine (AMP) and guanine (GMP). The absorption and emission peak position do not change in presence of either of these two. However, the fluorescence intensity of luminol quenches in presence of both AMP and GMP (Figure 12). A strong interaction of LH_2 with AMP and GMP is proposed based on the rapid quenching of luminol in presence of increasing concentrations of AMP and/or GMP. In the SV plot for the luminol fluorescence quenching (Figure 12), a downward curvature is observed. Apparent values of K_a and the fraction of fluorophore accessible to the quencher (f_a) can be obtained from the slope and intercept of the plot of (F_0/ ΔF) vs $[Q]^{-1}$ using Eq. (8). Linear variation of this modified SV plot is obtained for both AMP and GMP with acceptable statistical parameters. The fractional accessibility (f_a) of fluorophore (LH_2) for quenching with AMP and GMP is 55% and 95%, respectively. The binding constant (K_a) of the accessible fraction for AMP is 1.33(± 0.08) × 10^4 M^{-1}; whereas for GMP, the value is 13.67(±1.09) ×10^4 M^{-1}. These values further affirm a higher interaction of LH_2 with GMP than with AMP.

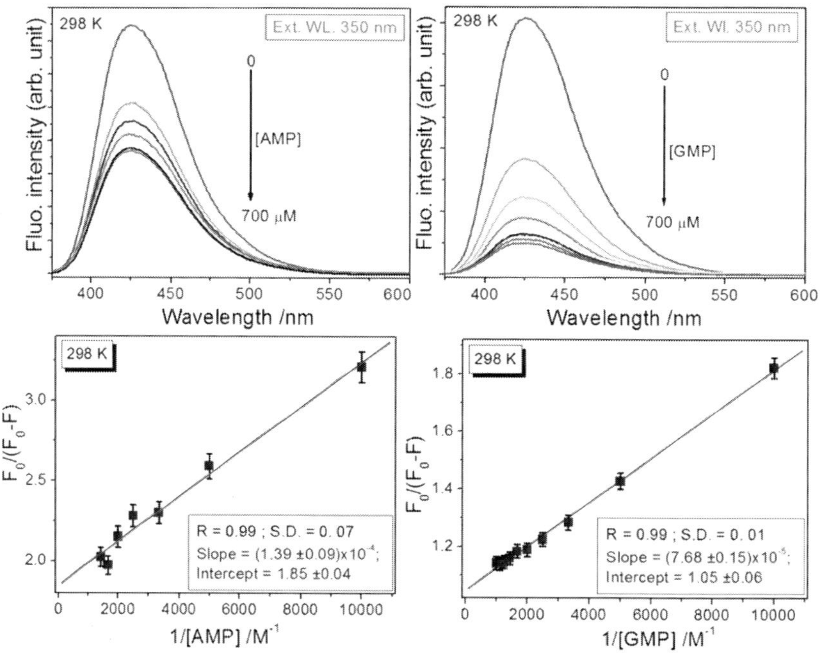

Figure 12. Fluorescence emission spectra of Luminol in presence of varying concentrations of AMP and GMP and modified SV plot for binding of LH_2 with AMP and GMP.

Time resolved fluorescence studies are also carried out to check the binding of LH_2 with AMP and GMP with varying quencher concentrations. Complete analysis of the decay data along with the average lifetime (calculated by using Eq. 4) values are depicted in Table 7. The lifetime of LH_2 decay does not change either in presence of AMP or GMP. Hence, we infer that the fluorescence quenching is static in nature. The association constants calculated from Eq. (9) render the values 19.5(±0.17) and 2.29

(±0.77)×10^3 M^{-1} for AMP and GMP, respectively confirming comparatively stronger interaction in the later case.

Table 7. Measurement of LH_2 fluorescence lifetime with varying concentrations of AMP and GMP[a]

[AMP] and/or [GMP] /mM	LH_2-AMP					LH_2-GMP				
	$\alpha_1(\%)$	τ_1 /ns	$\alpha_2(\%)$	τ_2 /ns	$<\tau>$ /ns	$\alpha_1(\%)$	τ_1 /ns	$\alpha_2(\%)$	τ_2 /ns	$<\tau>$ /ns
0.0	41	3.0	59	9.4	8.2	41	3.0	56	9.4	8.2
0.1	40	3.1	60	9.8	8.6	39	3.4	61	10.3	9.1
0.2	39	3.2	61	9.9	8.8	35	3.2	65	10.2	9.2
0.3	36	3.1	64	9.9	8.9	33	2.7	67	10.0	9.1
0.4	38	3.2	62	10.0	8.9	33	2.9	67	10.1	9.2
0.5	36	3.3	64	9.9	8.9	34	2.8	66	10.2	9.3
0.6	37	3.0	63	10.0	9.0	33	2.7	67	10.0	9.1
0.7	38	3.1	62	9.9	8.8	31	2.9	69	10.0	9.2
0.8	40	3.0	60	9.8	8.6	34	2.5	66	10.0	9.1
0.9	39	2.9	61	9.8	8.7	34	2.7	66	10.0	9.1
1.0	40	3.1	60	10.0	8.8	35	2.6	65	10.0	9.1

[a] Each fluorescence decay curves were analyzed with bi-exponential decay function using Eq. (3) and average lifetime are calculated by using Eq. (4).

3.7. Luminol Fluorescence Behavior at Nano-Bio Interface: An Efficient Model for Energy Transfer

Surface modified spherical metal nanoparticles (MNP) are known to act as significant energy acceptors for several excited organic molecules. Literature reports are available on the modification of MNP surface due to the adsorption of plasma proteins on it (Hao et al. 2017). Adsorption of bovine serum albumin (BSA) on different NP surface as a probe is a topic of immense interest because of its easy availability in pure form and extraordinary solubility in water. It is shown in the previous section that the fluorescence behavior of LH_2 can be strongly modulated in presence of a series of sequestering agents including BSA. These results prompt us to undertake a detailed study to verify the modulatory effect of metal NPs both in presence and absence of BSA on LH_2 luminescence.

3.7.1. Characterizing LH_2 – NP Interaction in Aqueous Buffer Medium

Both in presence of Ag and Au NPs, extremely efficient fluorescence quenching of LH_2 is observed (Figure 13i). It is interesting to note that the SV analysis of the quenching data gives different results for the investigated NPs; while a straight line is observed for AuNP, it shows an upward curvature for AgNP. Time-resolved fluorescence decay is also measured to confirm the quenching mechanism. On consecutive addition of the NPs, the fluorescence decay pattern of LH_2 does not give any meaningful change (inset, Figure 13i) and is found to remain bi-exponential (Figure 14) with similar average lifetime, indicating a static process in both the cases.

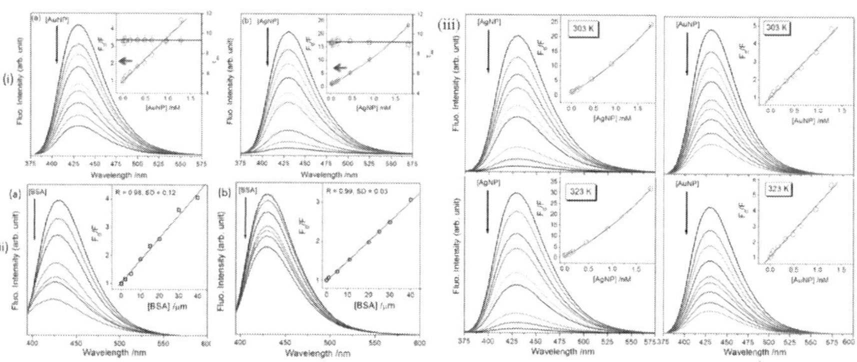

Figure 13. (i) LH_2 fluorescence quenching in presence increasing concentration of Au (a) and Ag (b) NPs. Inset shows the variation in F_0/F (left) or average lifetime (τ_{av}, right) obtained by using Eq. 4 with quencher concentration. (ii) Variation of LH_2 – NP composite fluorescence with the change in concentration of BSA. $[LH_2] = 1$ μM, [AgNP] = [AuNP] = 0.32 nM (b), [BSA] = 0 ~ 40 μM (along the arrow direction) in each case. Inset shows the SV plot. (iii) Quenching of LH_2 fluorescence at representative temperatures in presence different concentration of Au (right) and Ag (left) NPs. Inset shows the difference in Stern-Volmer analysis for the two cases.

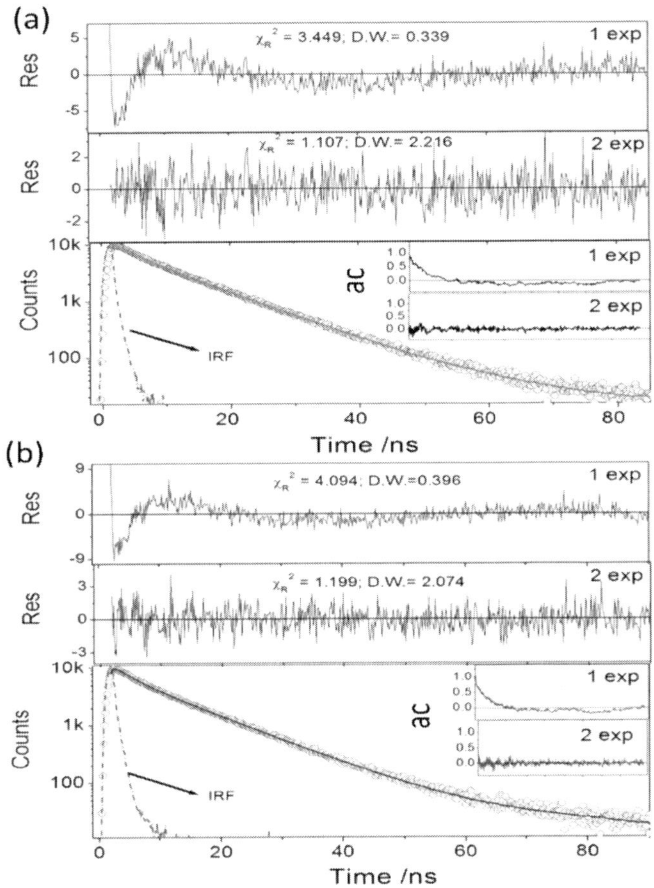

Figure 14. Time-resolved fluorescence decay (scattered points) data of LH_2 in presence of 0.96 nM AuNP (a) and 0.90 nM AgNP (b) showing 2-exponential decay function along with the simulation data (solid line). IRF is the instrument response function. Distribution of weighted residual (Res) and autocorrelation function (ac) confirms the acceptability of bi-exponential decay model.

The constancy of average lifetime rules out the probability of any dynamic or diffusive quenching mechanism. In case of AgNP, a positive deviation in SV plot indicates a situation where the extent of quenching is extremely high. The most common hypothesis for this is the development of a quenching sphere (Eftink and Ghiron 1976, Jin et al. 2009, Castanho and Prieto 1998), inside which the quenching probability is almost equal to unity due to efficient non-radiative energy transfer between the donor (excited

LH_2) and acceptor (AgNP) (see below). The factors which determine the possibility for the quencher to be inside this sphere at the instant of excitation are the volume (V) and the quencher concentration, [Q]. The linear SV equation is thereby modified to include the overall probability of this extra quenching, which gives the following relations.

$$\frac{F_0}{F} = \{1 + K_{SV}[Q]\} \times e^{V_q \cdot [Q]} \tag{13}$$

$$V_q = N_A \times V = N_A \times \frac{4}{3}\pi R_S^3 \tag{14}$$

Here, K_{SV} denotes the quenching constant that defines the linear part of the SV plot and R_S represents the radius of the quenching sphere.

Table 8 depicts the simulation results of LH_2 fluorescence quenching data on addition of Ag and AuNP at some representative temperatures. Apparently, the magnitude of K_{SV} is about 2~3 times higher for AgNP when compared with AuNP. The K_{SV} values, obtained by performing the quenching experiments at different temperatures, are used in the estimation of different thermodynamic parameters of LH_2-NP interaction. Negative values are obtained for ΔG, which further asserts that the binding process is spontaneous within the entire range of temperatures. Also, more negative ΔG value, attained for LH_2-AgNP system in contrast with LH_2-AuNP complex, corroborates well with the assumption of stronger interaction in the former (see below). In either of these cases, the magnitude of ΔH is found to be positive. Hence, the spontaneous adsorption process can be hypothesized to be primarily governed by overall positive entropy change. Due to the intrinsic complexity of the investigated systems and a profound difference in the quenching processes for Au and Ag NP systems as discussed above, the calculation of the binding stoichiometry is not straightforward. Despite these limitations, analysis of LH_2 fluorescence quenching data through the double-log plot (Eq. 9), shown in Figure 15, reveals the apparent stoichiometry of binding to be 1:1 in case of LH_2–AgNP system; while, the corresponding ratio for LH_2–AuNP system is estimated to be 3:2. Higher values of K_{SV} along with a positive deviation in the SV plot is indicative of relatively stronger interaction in the former case. This inference is consistent with a higher degree of reactivity and can be hypothesized to be a consequence of excess charge density (smaller size) of AgNP in comparison with AuNP (Su et al. 2014).

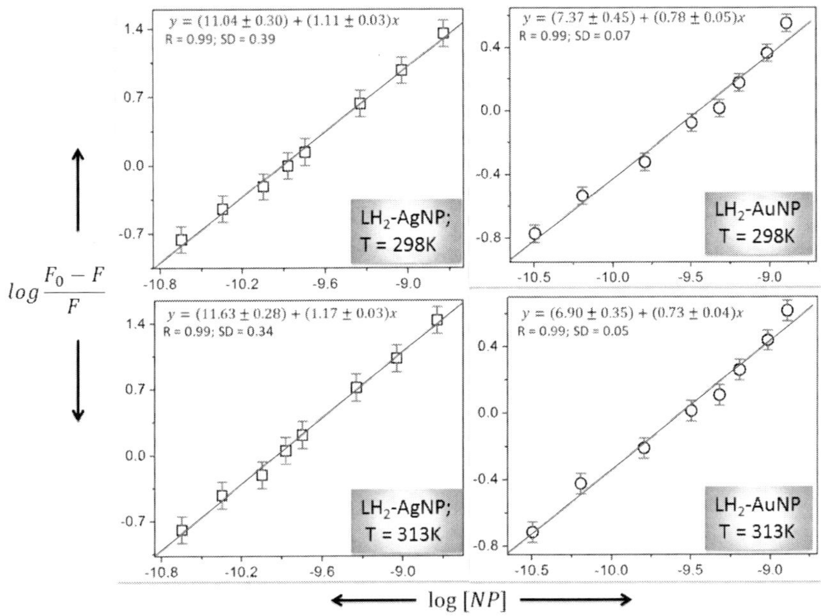

Figure 15. Double-log plot for LH_2 – NP interaction at different temperatures. Scattered points are the experimental data and solid line is the simulation result along with the correlation coefficient (R) and standard deviation (SD) in each case.

As already mentioned, the fluorescence decay data of LH_2 remains bi-exponential within the entire concentration range of added NPs. However, global analysis of the whole data set in presence of NPs exhibits interesting observations regarding the fractional contribution of each component (Table 9). In both the cases, the contribution of fast decay component suffers a significant decrease (~25%) with addition of NPs. It indicates that LH_2 preferentially binds with MNP surfaces with structure **I** and causes an overall diminution in emission intensity (quenching). The preference of binding of the NPs for conformer **I** seems validated when the basic structure of this conformer is compared with its counterpart, **II** (Figure 1). Conformer

I contain two adjacent amino groups which are suitably located to make a tweezer shaped anchor on the spherical NP surface when compared with isomeric structure **II** containing asymmetric distribution of binding regions.

Table 8. Thermodynamic and other binding parameters corresponding to the interaction of LH_2 with Ag and AuNP[a]

Temp (K)	AuNP				AgNP					
	K_{SV} /10^9, M^{-1}	ΔH	ΔS	ΔG	K_{SV} /10^9, M^{-1}	ΔH	ΔS	ΔG	V_q /10^8, cm^3	R_S, nm
298	2.56 ±0.12	8.16	207.60	-53.70	8.50 ±0.22	9.60	222.24	-56.62	2.01 ±0.13	43.0
303	2.77 ±0.11			-54.74	9.05 ±0.25			-57.73	1.87 ±0.14	42.0
308	2.88 ±0.14			-55.78	9.56 ±0.26			-58.85	1.85 ±0.14	41.8
313	2.98 ±0.13			-56.82	9.47 ±0.42			-59.96	2.51 ±0.23	46.3
318	3.15 ±0.14			-57.86	10.69 ±0.14			-61.07	2.35 ±0.24	45.3
323	3.36 ±0.15			-58.89	11.75 ±0.50			-62.18	2.07 ±0.22	43.4

[a]K_{SV} values are obtained from Eq. (7) and (8) for AuNp and AgNP, respectively, the error limit indicates the confidence interval in three independent measurements (mean±S.D.). ΔH and ΔG are represented in kJ mol^{-1}; whereas, ΔS is in J K^{-1} mol^{-1}.

3.7.2. Binding of LH_2 – NP Conjugate with Bovine Serum Albumin (BSA)

Fluorescence intensity of LH_2 adsorbed on MNP surface (hereinafter referred as LH_2 – NP conjugate) quenches regularly in presence of BSA (Figure 13ii), quite like the case of the homogeneous LH_2 solution discussed above. However, in contrast to the free LH_2 fluorescence, where a spectral shift of ~10 nm occurs, LH_2 – NP conjugate exhibits a blue shift of only around 5 and 3 nm for AgNP and AuNP, respectively on binding to BSA. From this observation it can be argued that the association of LH_2 adsorbed on NP surface within the protein binding domain (PBD) is less efficient than in the case of free LH_2. LH_2 – NP fluorescence quenching is represented by linear SV relationship (inset of Figure 13(ii)), while corresponding parameters have been listed in Table 10. The magnitude of K_{SV} is higher for

LH_2 – AgNP than the LH_2 - AuNP system. In both the cases however, the value is seen to be about an order of magnitude less in comparison to that of quenching of free LH_2 fluorescence by BSA.

As seen in the earlier section, LH_2 interacts with two PBDs of BSA (subdomains IA and IIA) with n_H and $K_{0.5}$ values of 2.2 and 6.0 μM, respectively. The n_H values for the interaction of LH_2 – NP conjugates in presence of BSA decrease significantly (ca. 1.3 ~ 1.7) at different temperatures (Figure 16i and Table 10). Also, the magnitude of $K_{0.5}$ is quite high (7~10 μM) in the present case. These results point towards relatively less efficient binding of protein with LH_2 – NP composite system. This further signifies the relevance and importance of nano-toxicity on serum albumin activity as carrier proteins reported in the literature (Saptarshi, Duschl, and Lopata 2013, De Jong and Borm 2008).

Table 9. Global analysis data of time-resolved fluorescence decay traces for different LH_2 –NP conjugate systems[a]

LH_2 – AgNP τ_1 = 2.96 ns; τ_2 = 10.14 ns				**LH_2 – AuNP** τ_1 = 3.17 ns; τ_2 = 10.28 ns			
[AgNP] /nM	α_1	α_2	τ_{av} /ns	[AuNP] /nM	α_1	α_2	τ_{av} /ns
0	38.22	61.78	9.04	0	39.17	60.83	8.80
0.02	36.41	63.59	9.11	0.03	36.59	63.41	8.91
0.04	36.11	63.89	9.12	0.06	35.64	64.36	8.95
0.09	35.02	64.98	9.16	0.16	34.47	65.53	8.99
0.13	33.46	66.54	9.22	0.32	33.32	66.68	9.04
0.18	32.21	67.79	9.27	0.48	33.06	66.94	9.04
0.45	30.87	69.13	9.31	0.64	31.99	68.01	9.08
0.90	31.25	68.75	9.30	0.96	30.9	69.1	9.12
1.80	31.04	68.96	9.31	1.28	29.85	70.15	9.16

[a] Average life time in different systems were calculated using Eq. (4).

Certain interesting features are observed in the TCSPC data analyses of LH_2-NP-BSA ternary system (Figure 17). The contribution of component **I** increases by about 25% when the concentration of protein is ca. 40 μM, probably due to the disruption of the weak **I** – AuNP interaction, and corresponding release of certain fraction of **I** in the solution mixture. The situation, however is far more complex and interesting in case of AgNP.

Table 10. Thermodynamic and other binding parameters corresponding to the interaction of LH_2 – NP composite with BSA[a]

T (K)	AuNP					AgNP				
	K_{SV} /10^4, M^{-1}	ΔH or ΔS	ΔG	n_H	$K_{0.5}$μM	K_{SV} /10^4, M^{-1}	ΔH or ΔS	ΔG	n_H	$K_{0.5}$, μM
298	5.10 ±0.07	ΔH = -6.26,	-26.91	1.61	9.66	7.89 ±0.64	ΔH = -33.66,	-28.31	1.62	6.78
303	5.02 ±0.08	ΔS = 69.27	-27.25	1.65	9.63	7.42 ±0.59	ΔS = -17.98	-28.22	1.63	7.10
308	4.84 ±0.06		-27.59	1.64	9.29	6.63 ±0.53		-28.13	1.51	7.04
313	4.66 ±0.72		-27.94	1.48	8.38	5.49 ±0.46		-28.04	1.59	8.22
318	4.42 ±0.81		-28.29	1.46	8.06	4.09 ±0.37		-27.95	1.71	9.82
323	4.21 ±0.09		-28.64	1.49	7.54	2.66 ±0.28		-27.86	2.23	10.00

[a]K_{SV} values are obtained from the linear regression of SV data (Eq. 7). The error limit indicates the confidence interval in three independent measurements (mean±S.D.). ΔH and ΔG are represented in kJ mol^{-1}; whereas, ΔS is in J K^{-1} mol^{-1}.

An additional decay parameter of 0.67 ns also appears with considerable contribution (>10%) along with the two decay components, even in presence of minimal amount of BSA (1 μM). The contribution of this component is observed to show a continuous increase until it reaches ca. 80%, when the BSA concentration is 40 μM (Figure 17b). The existence of a fast instrument limited (<100 ps) decay component for LH_2 in presence of BSA is explained earlier as due to the development of a "fluorophore-quencher" pair. An analogous process can also be envisaged for LH_2 - AgNP system, which shows a characteristic decay time of 0.67 ns. However, the absence of any dark complex formation in case of LH_2 – AuNP binary system reveals that either or both fluorophore/quencher pair is shielded from each other, while being adsorbed on MNP surface. This phenomenon results in lesser interaction between them. This observation is further supported by the comparatively lower value of K_{SV} and the lesser magnitude of fluorescence spectral blue shift of LH_2 – AuNP system in presence of BSA in contrast with LH_2 – AgNP conjugate discussed above.

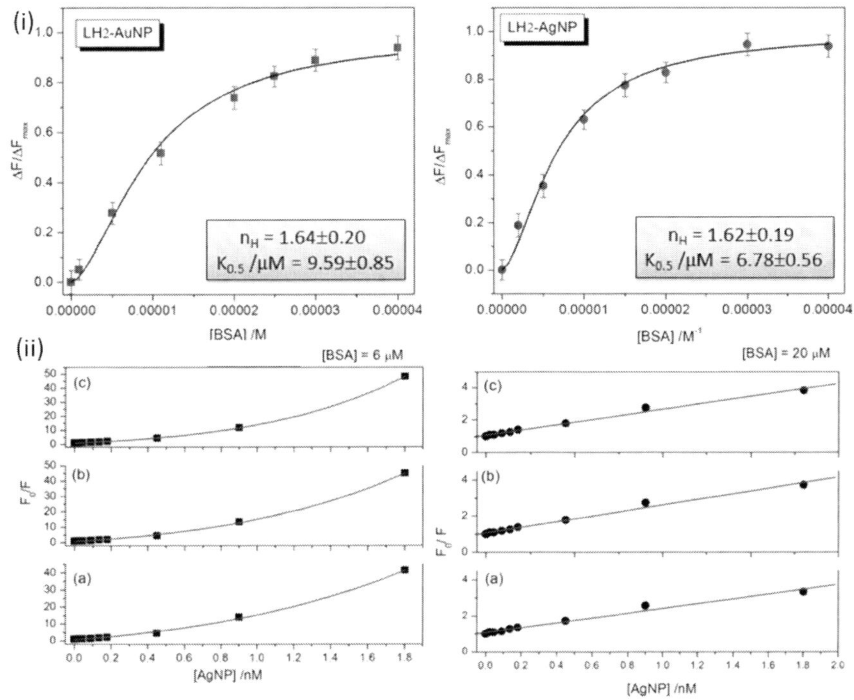

Figure 16. (i) Hill plots for LH_2 – NP – BSA system at 298 K. Simulation of experimental data points (solid line) are done using Eq. (12). (ii) SV analysis for the fluorescence quenching data of LH_2 – BSA assembly in presence of AgNP at temperature 298 K (a), 308 K (b) and 318 K (c). The solid lines represent the simulation of experimental data at [BSA] = 6 (square) and 20 μM (circle) by using equations (7) and (13), respectively.

The LH_2–NP composite system, which suffers a significant quenching in presence of serum albumin, can be represented by one site binding model ($n = 0.9 \pm 0.1$) containing comparable magnitude of ΔG values (Table 10) for both Ag and AuNP. However, while LH_2 – AuNP is seen to have a negative value of ΔH and positive value of ΔS, the binding of LH_2 – AgNP in BSA is associated with negative entropy value.

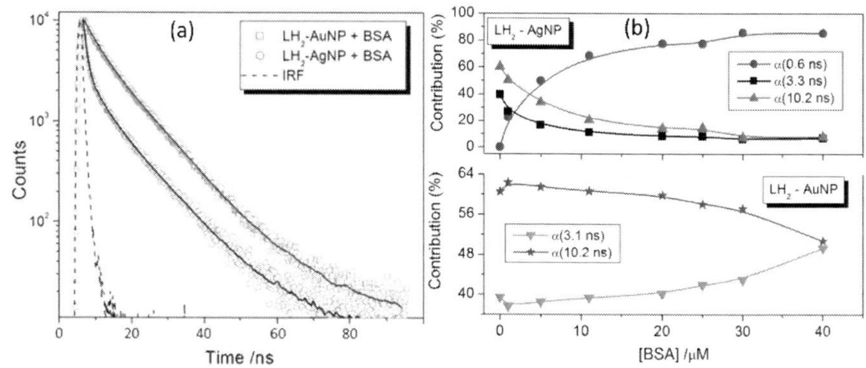

Figure 17. (a) Time-resolved fluorescence decay (scattered points) and simulated data (solid line) of LH_2 – NP composite system in presence of 20 μM BSA. IRF is the instrument response function shown by dashed line. (b) Fractional contribution (%) of different lifetime components in the fluorescence decay of LH_2 – NP assemblies with increase in the concentration of BSA.

From these observations, it can be concluded that both van der Waals interaction and hydrogen bond formation play a key role in the overall enthalpy driven process. These results are suggestive of the fact that LH_2 adsorbed on different metal surfaces have different interaction with BSA. The differences in the physical, chemical and optical properties of organic fluorophores adsorbed on Ag and AuNP surfaces are well documented in literature (Ansar et al. 2013). As an example, aromatic organo-thiol (R-S) compounds are known to react with AgNP forming deprotonated core-shell RS-Ag complex presumably due to its preferential interaction with silver oxide, which is either present or generated *in-situ* on AgNP surface. The interaction mechanism of these compounds varies significantly during adsorption on AgNP surface in comparison with AuNP (Gadogbe et al. 2014, Herrera, Padilla, and Hernandez-Rivera 2013). The differences observed in the experimental results in case of AgNP from the relatively idealized AuNP surface may be explained by a similar behavior.

3.7.3. Quenching of LH_2 Fluorescence Bound Inside BSA Cavity by MNPs

It is interesting to verify the interaction between LH_2 – BSA assembly with spherical MNPs. The modulation in fluorescence of BSA incubated LH_2 is checked on successive addition of both Ag and AuNP, keeping the

BSA concentration fixed either at ~6 or ~20 μM. The intensity of fluorescence is found to decrease regularly with continuous addition of the NPs. For the quenching of LH_2 with AgNP, the analysis of the quenching data is found to be consistent with linear S-V plot in case of 20 μM BSA as shown in Figure 16(ii); whereas, a sharp upward curvature is observed when the concentration of the BSA is kept at 6 μM. All the characteristic data for each of the systems obtained from the SV analysis are shown in Tables 11 and 12. The calculated thermodynamic parameters in different LH_2 – BSA – NP systems suggest that interaction is mainly due to hydrophobic mode of binding in all the cases. For AgNP in presence of 6 μM BSA, the structure of the protein environment is distorted along with the positive entropy change for binding, which suggest the importance of hydrogen bond formation as well as van der Waals interaction. Figure 18 shows the difference in the van't Hoff plot for the interaction of AgNP with aqueous LH_2 and LH_2 – BSA assembly, which signifies the specific role of BSA in the interaction process.

Quenching sphere model analysis of fluorescence intensity diminution data for LH_2 – BSA (6 μM) system in presence of AgNP gives certain interesting observations. The calculated values of R_S are found to be ~ 43±3 nm (in absence of albumin) at different temperatures for the case of AgNP; while the corresponding value is ca. ~ 65±5 nm when BSA (6 μM) is present. Therefore, it is believed that there is a specific role of BSA in bringing the fluorophore quencher pair in close vicinity of each other. When the concentration of the BSA is changed to 20 μM, no additional quenching phenomenon is observed (except normal static quenching process resulting in linear SV plot), which suggest that the stoichiometry of LH_2 – BSA – AgNP assembly system plays a critical role in the interaction process (see below).

Time resolved measurements are also monitored during successive addition of NPs into LH_2 - BSA system. The results are found to be similar as described in the previous section. In brief, during the quenching of LH_2 – BSA bio-conjugate system, the presence of an additional fast decay component of 0.5±0.1 ns is observed only in AgNP (at both the protein concentrations); whereas, a bi-exponential decay is observed always in case of the AuNP (Table 13).

Table 11. Thermodynamic and other binding parameters corresponding to the interaction of LH_2 incubated in 6 μM BSA with Ag and AuNP[a]

T (K)	AuNP				AgNP					
	K_{SV} /10^9, M^{-1}	ΔH	ΔS	ΔG	K_{SV} /10^9, M^{-1}	ΔH	ΔS	ΔG	V_q /10^8, cm^3	R_S, nm
298	1.61±0.11	10.84	212.70	-52.55	8.00±0.90	-20.9	119.71	-56.55	5.50±0.60	60.2
303	1.76±0.11			-53.61	7.14±0.70			-57.15	6.36±0.54	63.2
308	1.86±0.11			-54.67	6.74±0.68			-57.75	6.87±0.51	64.9
313	2.00±0.10			-55.74	5.09±0.31			-58.35	8.60±0.30	69.9
318	2.12±0.09			-56.80	4.59±0.18			-58.95	9.18±0.19	71.4
323	2.28±0.10			-57.86	4.43±0.16			-59.54	9.24±0.18	71.5

[a]K_{SV} values are obtained from Eq. (2a) and (7) for AuNp and AgNP, respectively. The error limit indicates the confidence interval in three independent measurements (mean±S.D.). ΔH and ΔG are represented in kJ mol^{-1}; whereas, ΔS is in J K^{-1} mol^{-1}.

Table 12. Thermodynamic and other binding parameters corresponding to the interaction of LH_2 incubated in 20 μM BSA with Ag and AuNP[a]

Temp (K)	AuNP				AgNP			
	K_{SV} /10^9, M^{-1}	ΔH	ΔS	ΔG	K_{SV} /10^9, M^{-1}	ΔH	ΔS	ΔG
298	1.52±0.11	6.06	196.19	-52.41	1.34±0.08	5.78	194.49	-52.17
303	1.60±0.12			-53.39	1.47±0.07			-53.15
308	1.65±0.13			-54.37	1.55±0.07			-54.12
313	1.72±0.12			-55.35	1.61±0.06			-55.09
318	1.79±0.12			-56.33	1.60±0.07			-56.06
323	1.85±0.12			-57.31	1.62±0.07			-57.04

[a]K_{SV} values are obtained from Eq. (7) and (8) for AuNp and AgNP, respectively. The error limit indicates the confidence interval in three independent measurements (mean±S.D.). ΔH and ΔG are represented in kJ mol^{-1}; whereas, ΔS is in J K^{-1} mol^{-1}.

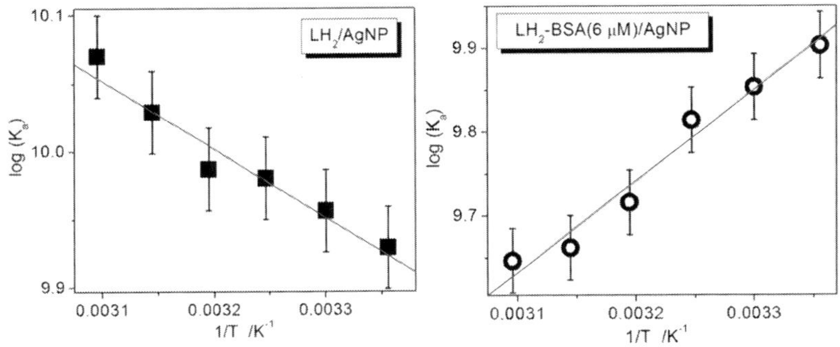

Figure 18. van't Hoff plot for the interaction of AgNP with aqueous LH_2 (left) and LH_2 – BSA (6 μM, right) assembly system.

Table 13. Global analysis data of time-resolved fluorescence decay traces for different LH_2 incubated with 6 μM BSA in presence of varying concentration of NPs

LH_2 – AgNP τ_1 = 0.41 ns; τ_2 = 3.15 ns; τ_3 = 10.33 ns				LH_2 – AuNP τ_1 = 3.21 ns; τ_2 = 10.11 ns		
[AgNP] /nM	α_1	α_2	α_3	[AuNP] /nM	α_1	α_2
0	26.60	24.98	48.42	0	51.92	48.08
0.02	28.92	23.73	47.35	0.06	49.50	50.50
0.04	27.63	24.16	48.21	0.16	48.94	51.06
0.09	36.78	20.09	43.13	0.32	44.73	55.27
0.13	39.14	19.68	41.18	0.48	42.60	57.40
0.18	43.02	18.07	38.91	0.64	40.91	59.09
0.45	58.06	13.80	28.14	0.96	38.07	61.93
0.90	80.29	7.32	12.38	1.41	35.24	64.76
1.80	90.75	4.12	5.13			

3.7.4. LH_2 – NP Assembly as a Super-Efficient Energy Transfer System in Presence of BSA

Large K_{SV} values signify that there is an efficient non-radiative energy transfer (ET) from LH_2 to the MNPs (Ghosh and Chattopadhyay 2013) both in absence or presence of BSA (Tables 8 and 11). Since the size of the synthesized NPs is extremely small (8~13 nm), the quenching phenomenon predominates over plasmonic near-field augmentation (Haes et al. 2004, Mirkin et al. 1996).

In the framework of Förster's theory (Andrews and Demidov 1999), the distance (r) between the donor (D) – acceptor (A) pair is related to the dipole induced ET efficiency (E) as

$$E = 1 - \frac{F}{F_0} = \frac{1}{1+(\frac{r}{R_0})^6} \quad (15)$$

R_0 is the critical Förster distance, defined as the distance where 50% energy transfer efficiency is achieved, and can be expressed as

$$R_0(\AA) = 0.211[\kappa^2 n^{-4} \phi_D J(\lambda)]^{1/6} \quad (16)$$

where, n is the refractive index of the medium with a value of 1.36 for aqueous buffer solution (Bogdan et al. 2008). Quantum yield (ϕ_D) of donor LH_2 is calculated to be 0.78 in aqueous buffer; whereas, in presence of 6 and 20 μM BSA the fluorescence quantum yields are 0.59 and 0.32, respectively. The degree of spectral overlap is represented by the overlap integral $J(\lambda)$ (shown in Figure 19) between D and A molecules, and can be expressed as

$$J(\lambda) = \frac{\int_0^\alpha F_D(\lambda)\varepsilon_A(\lambda)\lambda^4 d\lambda}{\int_0^\alpha F_D(\lambda) d\lambda} \quad (17)$$

The dimensionless quantity $F_D(\lambda)$ is the fluorescence intensity of D in the wavelength range of λ to $\lambda+d\lambda$, $\varepsilon_A(\lambda)$ is the molar absorptivity (in $dm^3\ mol^{-1}\ cm^{-1}$) of A at wavelength λ. In Eq. 16, κ^2 represent a factor describing the relative orientation of D and A; and usually taken as 2/3, considering the random distribution.

ET parameters are calculated for several systems and shown in Table 14. In all the cases, the average distance between the donor – acceptor pairs fall within the range of $0.5R_0 < r < 1.5R_0$, which suggest that there is a large degree of energy transfer between LH_2 donor and the NP acceptors (Mitchell et al. 2016, Ghosh, Girigoswami, and Chattopadhyay 2012). It is also very clear from Table 14 that the energy transfer efficiency is always more in case of AgNP system, even though the R_0 values are higher for AuNP. Therefore, it can be argued that although AuNP can induce as an acceptor even from a longer distance, AgNP forms more efficient donor-acceptor pair with LH_2 or hydrophobic assembly of LH_2 – BSA system. The more efficient energy transfer in the latter case is presumably due to high molar absorptivity of the acceptor. For spherical Ag and Au NPs with diameter 'd' (in nm), the molar absorptivity can be estimated from the following empirical relationship (Navarro and Werts 2013).

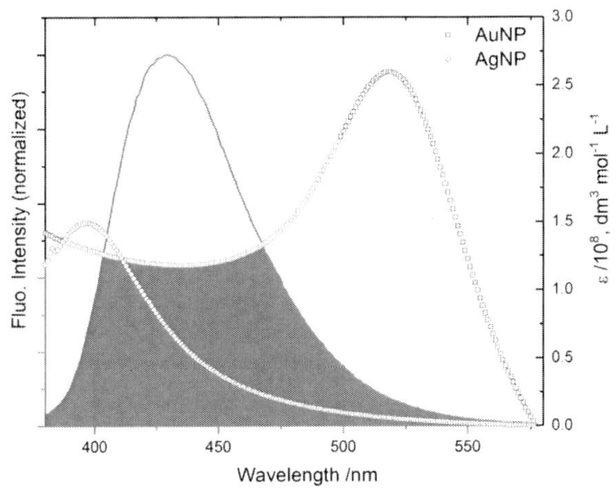

Figure 19. Spectral overlap (shaded portion) between the fluorescence (solid line) of LH_2 – BSA (6 μM) and the absorption of Au and Ag (scattered squares and circles, respectively) NP.

$$\epsilon_{max} = A \times d^{\gamma} \tag{18}$$

where 'A' and 'γ' are found to be ranging within 4.7×10^4 M^{-1} cm^{-1}, 3.30 and 2.3×10^5 M^{-1} cm^{-1}, 3.48 for AuNP (d≤85 nm) and AgNP (d≤38 nm), respectively. From the known size of the synthesized NPs, the calculated ε_{max} is indeed found to be at least one and half times higher for AgNP.

The value 'r' for different combination of FRET pairs is calculated and shown in Table 14. Careful observation of the calculated parameters signifies the importance of BSA in forming a super- efficient ET system.

Table 14. Calculated values of energy transfer efficiency (E), overlap integral (J), critical Förster's distance (R_0) and donor (D) – acceptor (A) distance (r) for several D – A combinations

Donor	Acceptor	E	J, /10^{18} dm^3 mol^{-1} cm^{-1} nm^4	R_0, /Å	r, /nm
LH_2	AuNP	0.78	9.579	224.74	18.20
LH_2	AgNP	0.54	2.116	174.74	17.06
LH_2 – BSA (6 μM)	AuNP	0.69	9.624	214.69	18.79
LH_2 – BSA (6 μM)	AgNP	0.97	2.128	166.95	9.35
LH_2 – BSA (20 μM)	AuNP	0.55	9.524	193.55	18.76
LH_2 – BSA (20 μM)	AgNP	0.66	2.185	151.43	13.59

The adsorption of LH_2 (donor) directly on the coated NP forms a simple D-spacer-A system with an average distance of 17±1 nm. However, in case of different NPs, the situation is more complex and works differently in the presence of BSA. On incubation, LH_2 preferentially binds to subdomain IIA of albumin through hydrophobic interaction. After the NPs are added to this assembly, a strong layer of "protein corona" is formed due to the adsorption of BSA on the NP surface (Rocker et al. 2009, Lacerda et al. 2010, Boulos et al. 2013). It is interesting to note that although the adsorption of BSA is "head-on" through a carboxylate – ammonia type salt bridge on citrate stabilized AuNP (Brewer et al. 2005), the interaction with AgNP is followed with a massive distortion in protein secondary structure (Dasgupta et al. 2016, De-Llanos et al. 2011) with possible hysteresis effect (Wang et al. 2017). Resonance light scattering experiment further suggests that adsorption of BSA on AgNP induce certain aggregated structure with increases in size than the individual BSA molecule (Mariam, Dongre, and Kothari 2011).

The observation discussed above can be justified with the difference in the structural behavior among different nano-bioconjugate systems. For example, the addition of AuNP in LH_2 – BSA system produces a soft protein corona with relatively longer D – A distance (with decreased ET efficiency) than the interaction between LH_2 – AgNP.

However, the proximity of D and A in BSA-AgNP system is because of the structural deformation and aggregation in BSA that result in increase in the ET efficiency. On the other hand, the drastic decrease in ET efficiency at higher BSA concentration (ca. 20 μM) even in presence of AgNP needs further justification. Literature reports suggest that the adsorption of BSA on AgNP surface gets saturated at [BSA]/[AgNP] ≈ 1 (Wang et al. 2017). The number of protein molecules adsorbed on the ~10 nm NP surface are roughly estimated to be 3.3×10^{12} molecules/cm^2 for "head on" adsorption and 2.0×10^{12} molecules/cm^2 for "side on" adsorption (in Au and AgNP, respectively) by considering the dimension of BSA as (5.5×5.5×9) nm (Yu et al. 2016, Rezwan et al. 2004). In the present experimental condition, the situations are arrived when the concentration of BSA is close to ca. 4.50 μM. However, at concentration near 20 μM, the ET efficiency between the donor (LH_2) and acceptor (AgNP) decreases, probably because of increased D – A separation due to excess BSA molecules in the solution.

In recent findings, ET from excited fluorophore adsorbed on MNP surface is termed as nanoparticle surface energy transfer (NSET) and propose r^{-4} dependence of ET efficiency. Using this model and Persson-Lang relation (Eq. 19), the value of R_0 can be calculated independently.

$$R_0 = \left(\frac{0.225c^3\phi_D}{\omega_D^2\omega_F\kappa_F}\right)^{1/4} \tag{19}$$

where, ϕ_D and ω_D are the fluorescence yield and angular frequency of D, ω_F and κ_F are the angular frequency and Fermi wave vector for the NPs. The values of ω_F and κ_F are known for the bulk gold (8.4×10^{15} s^{-1} and 1.2×10^8 cm^{-1}), respectively (Jennings, Singh, and Strouse 2006, Sen, Sadhu, and Patra 2007), the calculated R_0 values for LH_2 – AuNP and LH_2 – BSA (6 and 20 μM) – AuNP bio-conjugates systems are 8.66, 8.08 and 6.93 nm, respectively. These values are considerably less than even the size of the synthesized AuNP (13±2 nm). Therefore, the energy transfer in the above systems is more likely to occur via FRET rather than NSET.

CONCLUSION AND FUTURE PERSPECTIVES

Luminol (LH_2), widely popular for its intense chemiluminescence (CL), has found tremendous application in the broad areas of biology and chemistry encompassing the specific fields of forensic science as well as diagnostic and quality control. While the CL behavior of LH_2 has attracted considerable attention, not much of information is available on the fluorescence properties of it. However, fluorescence spectroscopy is known to be very sensitive and can be specifically used for diverse chemical and biological applications, provided the understanding of basic photophysical behavior of the concerned probe is clear. The results provided in the present chapter gives an in-depth information on the spectroscopic properties of LH_2 and its modulation in a variety of bio-mimicking environments paving the way for its use as a possible fluorescent solvatochromic probe and bio-marker. Extremely efficient energy transfer from excited LH_2 to the metal nanoparticles in serum albumin matrix can be exploited to develop a possible fluorescence nano-biosensor.

ACKNOWLEDGMENTS

Financial support through research projects 2009/37/26/BRNS and 01(2568)/12/EMR - II from the Board of Research in Nuclear Sciences (BRNS) and Council of Scientific and Industrial Research (CSIR), Government of India, is gratefully acknowledged. The authors also acknowledge the support from the Department of Science and Technology (DST), Govt. of India to the Chemistry Department through FIST program (SR/FST/CSI-194/2008). Computational facilities were provided by the Computer Center, NEHU.

REFERENCES

Abboud, Jose Luis, Mortimer J. Kamlet, and R. W. Taft. 1977. "Regarding a generalized scale of solvent polarities." *Journal of the American Chemical Society* 99 (25):8325-8327. doi: 10.1021/ja00467a039.

Agasti, Sarit S., Subinoy Rana, Myoung-Hwan Park, Chae Kyu Kim, Chang-Cheng You, and Vincent M. Rotello. 2010. "Nanoparticles for detection and diagnosis." *Advanced Drug Delivery Reviews* 62 (3):316-328. doi: doi.org/10.1016/j.addr.2009.11.004.

Andrews, D. L., and A. D. Demidov. 1999. *Resonance Energy Transfer*. England: John Wiley & Sons Ltd.

Ansar, Siyam M., Ganganath S. Perera, Patricia Gomez, George Salomon, Erick S. Vasquez, I. Wei Chu, Shengli Zou, Charles U. Pittman, Keisha B. Walters, and Dongmao Zhang. 2013. "Mechanistic Study of Continuous Reactive Aromatic Organothiol Adsorption onto Silver Nanoparticles." *The Journal of Physical Chemistry C* 117 (51):27146-27154. doi: 10.1021/jp4090102.

Banerjee, D., A. Mandal, and S. Mukherjee. 2003. "Excited state complex formation between methyl glyoxal and some aromatic bio-molecules: a fluorescence quenching study." *Spectrochimica Acta Part A: Molecular and Biomolecular Spectroscopy* 59 (1):103-109. doi: doi.org/10.1016/S1386-1425(02)00125-7.

Barni, F., S. W. Lewis, A. Berti, G. M. Miskelly, and G. Lago. 2007. "Forensic application of the luminol reaction as a presumptive test for latent blood detection." *Talanta* 72 (3):896-913. doi: 10.1016/j.talanta.2006.12.045.

Bevington, P.R., and D.K. Robinson. 2003. *Data Reduction and Error Analysis for the Physical Sciences*. 3rd Ed. ed. NY: McGraw-Hill.

Bi, S., H. Zhou, and S. Zhang. 2009. "Multilayers enzyme-coated carbon nanotubes as biolabel for ultrasensitive chemiluminescence immunoassay of cancer biomarker." *Biosensors and Bioelectronics* 24 (10):2961-2966. doi: doi.org/10.1016/j.bios.2009.03.002.

Bogdan, M., A. Pirnau, C. Floare, and C. Bugeac. 2008. "Binding interaction of indomethacin with human serum albumin." *J Pharm Biomed Anal* 47 (4-5):981-4. doi: 10.1016/j.jpba.2008.04.003.

Boulos, S. P., T. A. Davis, J. A. Yang, S. E. Lohse, A. M. Alkilany, L. A. Holland, and C. J. Murphy. 2013. "Nanoparticle-protein interactions: a thermodynamic and kinetic study of the adsorption of bovine serum albumin to gold nanoparticle surfaces." *Langmuir* 29 (48):14984-96. doi: 10.1021/la402920f.

Bowie, Andrew R., Matthew G. Sanders, and Paul J. Worsfold. 1996. "Analytical Applications of Liquid Phase Chemiluminescence Reactions — A Review." *Journal of Bioluminescence and Chemiluminescence* 11 (2):61-90. doi: doi:10.1002/(SICI)1099-1271.

Brewer, S. H., W. R. Glomm, M. C. Johnson, M. K. Knag, and S. Franzen. 2005. "Probing BSA binding to citrate-coated gold nanoparticles and surfaces." *Langmuir* 21 (20):9303-7. doi: 10.1021/la050588t.

Budavari, S. 1996. *The Merck Index (12th edition)*. USA: Merck & Co., Inc. Whitehall, NJ.

Carter, D. C., and X. M. He. 1990. "Structure of human serum albumin." *Science* 249 (4966):302-3.

Carter, D. C., and J. X. Ho. 1994. "Structure of serum albumin." *Adv Protein Chem* 45:153-203.

Castanho, M. A., and M. J. Prieto. 1998. "Fluorescence quenching data interpretation in biological systems. The use of microscopic models for data analysis and interpretation of complex systems." *Biochim Biophys Acta* 1373 (1):1-16.

Chatterjee, S., J. B. Lee, N. V. Valappil, D. Luo, and V. M. Menon. 2011. "Investigating the distance limit of a metal nanoparticle based spectroscopic ruler." *Biomedical optics express* 2 (6):1727-1733. doi: 10.1364/BOE.2.001727.

Chatterjee, T., D. Mukherjee, S. Dey, A. Pal, K. M. Hoque, and P. Chakrabarti. 2011. "Accessory cholera enterotoxin, Ace, from Vibrio cholerae: structure, unfolding, and virstatin binding." *Biochemistry* 50 (14):2962-72. doi: 10.1021/bi101673x.

Chen, H., F. Gao, R. He, and D. Cui. 2007. "Chemiluminescence of luminol catalyzed by silver nanoparticles." *J Colloid Interface Sci* 315 (1):158-63. doi: 10.1016/j.jcis.2007.06.052.

Chuang, V. T., and M. Otagiri. 2006. "Stereoselective binding of human serum albumin." *Chirality* 18 (3):159-66. doi: 10.1002/chir.20237.

Committee, on DNA Technology in Forensic Science. 1992. *DNA technology in forensic science*. Washington DC National Research Council, National Academy press

Connors, K. A. 1987. *Binding Constants. The Measurements of Molecular Complex Stability*. New York: Wiley.

Dasgupta, N., S. Ranjan, D. Patra, P. Srivastava, A. Kumar, and C. Ramalingam. 2016. "Bovine serum albumin interacts with silver nanoparticles with a "side-on" or "end on" conformation." *Chem Biol Interact* 253:100-11. doi: 10.1016/j.cbi.2016.05.018.

De-Llanos, R., S. Sánchez-Cortes, C. Domingo, J. V. García-Ramos, and P. Sevilla. 2011. "Surface Plasmon Effects on the Binding of Antitumoral Drug Emodin to Bovine Serum Albumin." *The Journal of Physical Chemistry C* 115 (25):12419-12429. doi: 10.1021/jp111683c.

De Jong, W. H., and P. J. Borm. 2008. "Drug delivery and nanoparticles:applications and hazards." *Int J Nanomedicine* 3 (2):133-49.

Dockal, M., D. C. Carter, and F. Ruker. 1999. "The three recombinant domains of human serum albumin. Structural characterization and ligand binding properties." *J Biol Chem* 274 (41):29303-10.

Du, J., Y. Li, and J. Lu. 2001. "Investigation on the chemiluminescence reaction of luminol–H2O2–S2−/R–SH system." *Analytica Chimica Acta* 448 (1):79-83. doi: doi.org/10.1016/S0003-2670(01)01242-9.

Eftink, M. R., and C. A. Ghiron. 1976. "Fluorescence quenching of indole and model micelle systems." *The Journal of Physical Chemistry* 80 (5):486-493. doi: 10.1021/j100546a014.

Foster, J. F., M. Sogami, H. A. Petersen, and W. J. Leonard, Jr. 1965. "The microheterogeneity of plasma albumins. I. Critical evidence for and description of the microheterogeneity model." *J Biol Chem* 240:2495-502.

Gadogbe, M., S. M. Ansar, I. W. Chu, S. Zou, and D. Zhang. 2014. "Comparative study of the self-assembly of gold and silver nanoparticles onto thiophene oil." *Langmuir* 30 (39):11520-7. doi: 10.1021/la502574p.

Ghosh, D., and N. Chattopadhyay. 2013. "Gold Nanoparticles: Acceptors for Efficient Energy Transfer from the Photoexcited Fluorophores." *Optics and Photonics Journal* Vol.03No.01:9. doi: 10.4236/opj.2013.31004.

Ghosh, D., A. Girigoswami, and N. Chattopadhyay. 2012. "Superquenching of coumarin 153 by gold nanoparticles." *Journal of Photochemistry and Photobiology A: Chemistry* 242:44-50. doi: doi.org/10.1016/j.jphotochem.2012.05.027.

Gross, A. M., K. A. Harris, and G. L. Kaldun. 1999. "The effect of luminol on presumptive tests and DNA analysis using the polymerase chain reaction." *J Forensic Sci* 44 (4):837-40.

Guha, D., U. Bhattacharjee, S. Mitra, R. Das, and S. Mukherjee. 1998. "Interaction of 3-aminophthalhydrazide with 5-hydroxytetracycline and chloramphenicol: a fluorescence quenching study." *Spectrochimica Acta Part A: Molecular and Biomolecular Spectroscopy* 54 (3):525-533. doi: doi.org/10.1016/S1386-1425(97)00244-8.

Guo, J-Z., H. Cui, W. Zhou, and W. Wang. 2008. "Ag nanoparticle-catalyzed chemiluminescent reaction between luminol and hydrogen peroxide." *Journal of Photochemistry and Photobiology A: Chemistry* 193 (2):89-96. doi: doi.org/10.1016/j.jphotochem.2007.04.034.

Haes, A. J., S. Zou, G. C. Schatz, and R. P. Van Duyne. 2004. "Nanoscale Optical Biosensor: Short Range Distance Dependence of the Localized Surface Plasmon Resonance of Noble Metal Nanoparticles." *The Journal of Physical Chemistry B* 108 (22):6961-6968. doi: 10.1021/jp036261n.

Haiss, W., N. T. K. Thanh, J. Aveyard, and D. G. Fernig. 2007. "Determination of Size and Concentration of Gold Nanoparticles from UV−Vis Spectra." *Analytical Chemistry* 79 (11):4215-4221. doi: 10.1021/ac0702084.

Hajime, K. 1987. "Microenvironmental Effects of Water-Soluble Polymers on the Chemiluminescence of Luminol and Its Analogs." *Bulletin of the Chemical Society of Japan* 60 (6):2023-2029. doi: 10.1246/bcsj.60.2023.

Hajipour, M. J., K. M. Fromm, A. A. Ashkarran, D. J. de Aberasturi, I. R. de Larramendi, T. Rojo, V. Serpooshan, W. J. Parak, and M. Mahmoudi. 2012. "Antibacterial properties of nanoparticles." *Trends in Biotechnology* 30 (10):499-511. doi: doi.org/10.1016/j.tibtech.2012.06.004.

Hao, C., G. Xu, Y. Feng, L. Lu, W. Sun, and R. Sun. 2017. "Fluorescence quenching study on the interaction of ferroferric oxide nanoparticles with bovine serum albumin." *Spectrochim Acta A Mol Biomol Spectrosc* 184:191-197. doi: 10.1016/j.saa.2017.05.004.

Hao, M-H. 2006. "Theoretical Calculation of Hydrogen-Bonding Strength for Drug Molecules." *Journal of Chemical Theory and Computation* 2 (3):863-872. doi: 10.1021/ct0600262.

He, S., W. Shi, X. Zhang, J. Li, and Y. Huang. 2010. "β−cyclodextrins-based inclusion complexes of CoFe2O4 magnetic nanoparticles as catalyst for the luminol chemiluminescence system and their applications in hydrogen peroxide detection." *Talanta* 82 (1):377-383. doi: doi.org/10.1016/j.talanta.2010.04.055.

He, X. M., and D. C. Carter. 1992. "Atomic structure and chemistry of human serum albumin." *Nature* 358 (6383):209-15. doi: 10.1038/358209a0.

Herrera, G. M., A. C. Padilla, and S. P. Hernandez-Rivera. 2013. "Surface Enhanced Raman Scattering (SERS) Studies of Gold and Silver Nanoparticles Prepared by Laser Ablation." *Nanomaterials (Basel)* 3 (1):158-172. doi: 10.3390/nano3010158.

Hoshino, H., and W. L. Hinze. 1987. "Exploitation of reversed micelles as a medium in analytical chemiluminescence measurements with application to the determination of hydrogen peroxide using luminol." *Analytical Chemistry* 59 (3):496-504. doi: 10.1021/ac00130a027.

Huang, X., and J. Ren. 2011. "Gold nanoparticles based chemiluminescent resonance energy transfer for immunoassay of alpha fetoprotein cancer marker." *Analytica Chimica Acta* 686 (1):115-120. doi: doi.org/ 10.1016/j.aca.2010.11.043.

Huang, X., and J. Ren. 2012. "Nanomaterial-based chemiluminescence resonance energy transfer: A strategy to develop new analytical methods." *TrAC Trends in Analytical Chemistry* 40:77-89. doi: doi.org/10.1016/j.trac.2012.07.014.

Huntress, E., L. Stanley, and A. Parker. 1934. "The Preparation of 3-Aminophthalhydrazide for Use in the Demonstration of Chemiluminescence." *Journal of the American Chemical Society* 56 (1):241-242. doi: 10.1021/ja01316a077.

Irie, S. 1960a. "Influence of 3-aminophthalhydrazide on the prothrombin time." *Curr Ther Res Clin Exp* 2:153-7.

Irie, S. 1960b. "Observations on the diuretic effects of 3-aminophthalhydrazide." *Curr Ther Res Clin Exp* 2:132-6.

Irie, S. 1960c. "The treatment of alopecia areata with 3-aminophthalhydrazide." *Curr Ther Res Clin Exp* 2:107-10.

Jennings, T. L., M. P. Singh, and G. F. Strouse. 2006. "Fluorescent Lifetime Quenching near d = 1.5 nm Gold Nanoparticles: Probing NSET Validity." *Journal of the American Chemical Society* 128 (16):5462-5467. doi: 10.1021/ja0583665.

Jin, B., K. S. Min, S. W. Han, and S. K. Kim. 2009. "DNA-binding geometry dependent energy transfer from 4',6-diamidino-2-phenylindole to cationic porphyrins." *Biophys Chem* 144 (1-2):38-45. doi: 10.1016/j.bpc.2009.06.003.

Jin, Y-J., Y.-J. Luo, G.-P. Li, J. Li, Y.-F. Wang, R.-Q. Yang, and W.-T. Lu. 2008. "Application of photoluminescent CdS/PAMAM nanocomposites in fingerprint detection." *Forensic Science International* 179 (1):34-38. doi: doi.org/10.1016/j.forsciint.2008.04.010.

Kahl, R., A. Weimann, S. Weinke, and A. G. Hildebrandt. 1987. "Detection of oxygen activation and determination of the activity of antioxidants towards reactive oxygen species by use of the chemiluminigenic probes luminol and lucigenin." *Archives of Toxicology* 60 (1):158-162. doi: 10.1007/BF00296971.

Kamlet, M. J., J. L. M. Abboud, and R. W. Taft. 1981. *Prog. Phys. Org. Chem.*:485-630.

Kamlet, M. J., J. L. M. Abboud, M. H. Abraham, and R. W. Taft. 1983. "Linear solvation energy relationships. 23. A comprehensive collection of the solvatochromic parameters, pi*, alpha, and beta, and some methods for simplifying the generalized solvatochromic equation." *The Journal of Organic Chemistry* 48 (17):2877-2887. doi: 10.1021/jo00165a018.

Karatani, H. 1986. "Effects of cyclodextrin on enhancement for chemiluminescence of the luminol related compound." *Chemistry Letters* 15 (3):377-380. doi: 10.1246/cl.1986.377.

Karube, I., K. Yano, S. Sasaki, Y. Nomura, and K. Ikebukuro. 1998. "Biosensors for environmental monitoring." *Ann N Y Acad Sci* 864:23-36.

Khan, P., D. Idrees, M. A. Moxley, J. A. Corbett, F. Ahmad, G. von Figura, W. S. Sly, A. Waheed, and M. I. Hassan. 2014. "Luminol-based chemiluminescent signals: clinical and non-clinical application and future uses." *Applied biochemistry and biotechnology* 173 (2):333-355. doi:10.1007/s12010-014-0850-1.

Kricka, L. J. 1995. "Chemiluminescence and bioluminescence." *Analytical Chemistry* 67 (12):499-502. doi: 10.1021/ac00108a035.

Lacerda, S. H., J. J. Park, C. Meuse, D. Pristinski, M. L. Becker, A. Karim, and J. F. Douglas. 2010. "Interaction of gold nanoparticles with common human blood proteins." *ACS Nano* 4 (1):365-79. doi: 10.1021/nn9011187.

Lakowicz, J. R. 2006. *Principles of fluorescence spectroscopy.* 3rd Ed. Singapore: Springer

Lasovský, J., and F. Grambal. 1986. "Micellar complexes of energy transfer." *Bioelectrochemistry and Bioenergetics* 15 (1):95-102. doi: doi.org/10.1016/0302-4598(86)80008-3.

Li, B., D. Wang, J. Lv, and Z. Zhang. 2006. "Flow-injection chemiluminescence simultaneous determination of cobalt(II) and copper(II) using partial least squares calibration." *Talanta* 69 (1):160-5. doi: 10.1016/j.talanta.2005.09.017.

Li, S.-F., X.-M. Zhang, W.-X. Du, Y.-H. Ni, and X.-W. Wei. 2009. "Chemiluminescence Reactions of a Luminol System Catalyzed by ZnO Nanoparticles." *The Journal of Physical Chemistry C* 113 (3):1046-1051. doi: 10.1021/jp808312j.

Li, X., H. Liu, X. He, and Z. Song. 2010. "Determination of cytochrome c in human serum and pharmaceutical injections using flow injection chemiluminescence." *Appl Biochem Biotechnol* 160 (4):1065-73. doi: 10.1007/s12010-009-8598-8.

Liu, C., and B. Li. 2011. "Silver nanoparticle-initiated chemiluminescence reaction of luminol-$AgNO_3$ and its analytical application." *Anal Bioanal Chem* 401 (1):229-35. doi: 10.1007/s00216-011-5071-7.

Maeztu, R., G. González-Gaitano, and G. Tardajos. 2010. "Enhancement of the Chemiluminescence of Two Isoluminol Derivatives by Nanoencapsulation with Natural Cyclodextrins." *The Journal of Physical Chemistry B* 114 (32):10541-10549. doi: 10.1021/jp103546u.

Maeztu, R., G. Tardajos, and G. González-Gaitano. 2010. "Natural Cyclodextrins as Efficient Boosters of the Chemiluminescence of Luminol and Isoluminol: Exploration of Potential Applications." *The Journal of Physical Chemistry B* 114 (8):2798-2806. doi: 10.1021/jp909707x.

Maleki, M. S., O. Moradi, and S. Tahmasebi. 2017. "Adsorption of albumin by gold nanoparticles: Equilibrium and thermodynamics studies." *Arabian Journal of Chemistry* 10:S491-S502. doi: doi.org/10.1016/j.arabjc.2012.10.009.

Marcus, Y. 1993. "The Properties of Organic Liquids that are Relevant to their Use as Solvating Solvents." *Chem. Soc. Rev.*:409-416.

Mariam, J., P. M. Dongre, and D. C. Kothari. 2011. "Study of Interaction of Silver Nanoparticles with Bovine Serum Albumin Using Fluorescence Spectroscopy." *Journal of Fluorescence* 21 (6):2193. doi: 10.1007/s10895-011-0922-3.

Mataga, N.; Kubota, T. 1970. *In Molecular interactions and electronic spectra*. New York: Marcel Dekker Inc.

Matsushita, Y., T. Suzuki, T. Ichimura, and T. Hikida. 2004. "Cavity Size Effect on the Excited State Dynamics of Methyl 4-(Dimethylamino)benzoate−Cyclodextrin Complexes." *The Journal of Physical Chemistry A* 108 (37):7490-7496. doi: 10.1021/jp036624j.

Miller, J. V., and E. G. Bartick. 2001. "Forensic Analysis of Single Fibers by Raman Spectroscopy." *Applied Spectroscopy* 55 (12):1729-1732. doi: 10.1366/0003702011954099.

Mirkin, C. A., R. L. Letsinger, R. C. Mucic, and J. J. Storhoff. 1996. "A DNA-based method for rationally assembling nanoparticles into macroscopic materials." *Nature* 382 (6592):607-9. doi: 10.1038/382607a0.

Mitchell, J. A., J. H. Whitfield, W. H. Zhang, C. Henneberger, H. Janovjak, M. L. O'Mara, and C. J. Jackson. 2016. "Rangefinder: A Semisynthetic FRET Sensor Design Algorithm." *ACS Sensors* 1 (11):1286-1290. doi: 10.1021/acssensors.
6b00576.

Mitra, S., R. Das, and S. Mukherjee. 1995. "Complex formation and photophysical properties of luminol: solvent effects." *Journal of Photochemistry and Photobiology A: Chemistry* 87 (3):225-230. doi: doi.org/10.1016/1010-6030(94)03995-7.

Moyon, N. S., A. K. Chandra, and S. Mitra. 2010. "Effect of Solvent Hydrogen Bonding on Excited-State Properties of Luminol: A Combined Fluorescence and DFT Study." *The Journal of Physical Chemistry A* 114 (1):60-67. doi: 10.1021/jp907970b.

Moyon, N. S., and S. Mitra. 2010. "On the interaction of luminol with human serum albumin: Nature and thermodynamics of ligand binding." *Chemical Physics Letters* 498 (1):178-183. doi: doi.org/10.1016/j.cplett.2010.08.029.

Nakamura, M., and S. Nakamura. 1998. "One- and Two-Electron Oxidations of Luminol by Peroxidase Systems." *Free Radical Biology and Medicine* 24 (4):537-544. doi: doi.org/10.1016/S0891-5849(97)00326-2.

National Research, Council. 1992. *DNA Technology in Forensic Science.* Washington, DC: The National Academies Press.

Navarro, J. R., and M. H. Werts. 2013. "Resonant light scattering spectroscopy of gold, silver and gold-silver alloy nanoparticles and optical detection in microfluidic channels." *Analyst* 138 (2):583-92. doi: 10.1039/c2an36135c.

Niemeyer, C. M. 2001. "Nanoparticles, Proteins, and Nucleic Acids: Biotechnology Meets Materials Science." *Angewandte Chemie International Edition* 40 (22):4128-4158. doi: 10.1002/1521-3773 (20011119)40:22<4128::AID-ANIE4128>3.0.CO;2-S.

Ogino, H. 1981. "Relatively high-yield syntheses of rotaxanes. Syntheses and properties of compounds consisting of cyclodextrins threaded by alpha, omega.-diaminoalkanes coordinated to cobalt(III) complexes." *Journal of the American Chemical Society* 103 (5):1303-1304. doi: 10.1021/ja00395a091.

Olson, R. E., and D. D. Christ. 1996. "Chapter 33. Plasma Protein Binding of Drugs." In *Annual Reports in Medicinal Chemistry*, edited by James A. Bristol, 327-336. Academic Press.

Peters, T., Jr. 1985. "Serum albumin." *Adv Protein Chem* 37:161-245.

Petersen, C. E., C. E. Ha, K. Harohalli, J. B. Feix, and N. V. Bhagavan. 2000. "A dynamic model for bilirubin binding to human serum albumin." *J Biol Chem* 275 (28):20985-95. doi: 10.1074/jbc.M001038200.

Pratap, P. R., L. O. Mikhaylyants, and N. Olden-Stahl. 2009. "Fluorescence measurements of nucleotide association with the Na(+)/K(+)-ATPase." *Biochim Biophys Acta* 1794 (11):1549-57. doi: 10.1016/j.bbapap.2009.06.023.

Rezwan, K., L. P. Meier, M. Rezwan, J. Vörös, M. Textor, and L. J. Gauckler. 2004. "Bovine Serum Albumin Adsorption onto Colloidal Al2O3 Particles: A New Model Based on Zeta Potential and UV−Vis Measurements." *Langmuir* 20 (23):10055-10061. doi: 10.1021/la048459k.

Rocker, C., M. Potzl, F. Zhang, W. J. Parak, and G. U. Nienhaus. 2009. "A quantitative fluorescence study of protein monolayer formation on colloidal nanoparticles." *Nat Nanotechnol* 4 (9):577-80. doi: 10.1038/nnano.2009.195.

Rosi, N. L., and C. A. Mirkin. 2005. "Nanostructures in Biodiagnostics." *Chemical Reviews* 105 (4):1547-1562. doi: 10.1021/cr030067f.

Ross, P. D., and S. Subramanian. 1981. "Thermodynamics of protein association reactions: forces contributing to stability." *Biochemistry* 20 (11):3096-3102. doi: 10.1021/bi00514a017.

Safavi, A., G. Absalan, and F. Bamdad. 2008. "Effect of gold nanoparticle as a novel nanocatalyst on luminol–hydrazine chemiluminescence system and its analytical application." *Analytica Chimica Acta* 610 (2):243-248. doi: doi.org/10.1016/j.aca.2008.01.053.

Saptarshi, S. R., A. Duschl, and A. L. Lopata. 2013. "Interaction of nanoparticles with proteins: relation to bio-reactivity of the nanoparticle." *J Nanobiotechnology* 11:26. doi: 10.1186/1477-3155-11-26.

Seitz, and W. Rudolf. 1975. "Chemiluminescence from the reaction between hypochlorite and luminol." *The Journal of Physical Chemistry* 79 (2):101-106. doi: 10.1021/j100569a002.

Sen, T., S. Sadhu, and A. Patra. 2007. "Surface energy transfer from rhodamine 6G to gold nanoparticles: A spectroscopic ruler." *Applied Physics Letters* 91 (4):043104. doi: 10.1063/1.2762283.

Shleev, S., J. Wetterö, K.-E. Magnusson, and T. Ruzgas. 2008. "Simultaneous use of electrochemistry and chemiluminescence to detect reactive oxygen species produced by human neutrophils." *Cell Biology International* 32 (12):1486-1496. doi: doi:10.1016/j.cellbi.2008.08.016.

Silbey, R. J., and R. A. Alberty. 2002. *Physical Chemistry (3rd Ed).* Singapore: John Wiley & Sons (Asia) Pte. Ltd.

Sonu, V. K., I. Rajkumar, K. Bhattacharjee, S. R. Joshi, and S. Mitra. 2019. "Interaction of caffeine and sulfadiazine with lysozyme adsorbed at colloidal metal nanoparticle interface: influence on drug transport ability and antibacterial activity." *Journal of Biomolecular Structure and Dynamics* 37:321-335. doi: 10.1080/07391102.2016.1192485.

Sonu, V. K., M. M. Islam, M. A. Rohman, and S. Mitra. 2016. "Lysozyme binding ability toward psychoactive stimulant drugs: Modulatory effect of colloidal metal nanoparticles." *Colloids and Surfaces B:*

Biointerfaces 146:514-522. doi: doi.org/10.1016/j.colsurfb.2016.06.061.

Stockwell, B. R., S. J. Haggarty, and S. L. Schreiber. 1999. "High-throughput screening of small molecules in miniaturized mammalian cell-based assays involving post-translational modifications." *Chem Biol* 6 (2):71-83. doi: 10.1016/s1074-5521(99)80004-0.

Su, Y., Y. Xie, X. Hou, and Y. Lv. 2014. "Recent Advances in Analytical Applications of Nanomaterials in Liquid-Phase Chemiluminescence." *Applied Spectroscopy Reviews* 49 (3):201-232. doi: 10.1080/05704928.2013.819514.

Sudlow, G., D. J. Birkett, and D. N. Wade. 1975. "The characterization of two specific drug binding sites on human serum albumin." *Mol Pharmacol* 11 (6):824-32.

Sudlow, G., D. J. Birkett, and D. N. Wade. 1976. "Further characterization of specific drug binding sites on human serum albumin." *Mol Pharmacol* 12 (6):1052-61.

Szejtli, J. 1985. "*In Controlled drug bioavailability*." V. Smolen (Ed), 365 – 420. NY: Wiley.

Takayuki, K., O. Minoru, Y. Masaaki, and S. Shigetaka. 1986. "Analytical Use of Luminescence Induced Ultrasonically in Solution. I. Sonic Chemiluminescence of Luminol for Determination of Cobalt(II) at Sub-pg Levels by Flow Injection and Continuous Flow Methods." *Bulletin of the Chemical Society of Japan* 59 (6):1849-1855. doi: 10.1246/bcsj.59.1849.

Tan, X., and Z. Song. 2014. "Human saliva-based quantitative monitoring of clarithromycin by flow injection chemiluminescence analysis: a pharmacokinetic study." *Appl Biochem Biotechnol* 172 (3):1320-31. doi: 10.1007/s12010-013-0605-4.

Triantis, T. M., K. Papadopoulos, E. Yannakopoulou, D. Dimotikali, J. Hrbáč, and R. Zbořil. 2008. "Sensitized chemiluminescence of luminol catalyzed by colloidal dispersions of nanometer-sized ferric oxides." *Chemical Engineering Journal* 144 (3):483-488. doi: 10.1016/j.cej.2008.07.035.

Tsach, T., S. Wiesner, and Y. Shor. 2007. "Empirical proof of physical match: Systematic research with tensile machine." *Forensic Science International* 166 (1):77-83. doi: doi.org/10.1016/j.forsciint.2006.04.002.

Tsogas, G. Z., D. L. Giokas, and A. G. Vlessidis. 2014. "Ultratrace Determination of Silver, Gold, and Iron Oxide Nanoparticles by Micelle Mediated Preconcentration/Selective Back-Extraction Coupled with Flow Injection Chemiluminescence Detection." *Analytical Chemistry* 86 (7):3484-3492. doi: 10.1021/ac404071v.

Upadhyayula, V. K. K. 2012. "Functionalized gold nanoparticle supported sensory mechanisms applied in detection of chemical and biological threat agents: A review." *Analytica Chimica Acta* 715:1-18. doi: doi.org/10.1016/j.aca.2011.12.008.

Valeur, B. 2002. *Molecular Fluorescence Principles and Applications*. Weinheim, FRG: Wiley-VCH.

Vasilescu, M., T. Constantinescu, M. Voicescu, H. Lemmetyinen, and E. Vuorimaa. 2003. *Journal of Fluorescence* "Spectrophotometric Study of Luminol in Dimethyl Sulfoxide–Potassium Hydroxide." 13 (4):315-322.

Velazquez, M. M., M. Valero, L. J. Rodriguez, S. M. Costa, and M. A. Santos. 1995. "Hydrogen bonding in a non-steroidal anti-inflammatory drug--Naproxen." *J Photochem Photobiol B* 29 (1):23-31.

Wang, G., Y. Lu, H. Hou, and Y. Liu. 2017. "Probing the binding behavior and kinetics of silver nanoparticles with bovine serum albumin." *RSC Advances* 7 (15):9393-9401. doi: 10.1039/C6RA26089F.

Wang, Z., N. Duan, J. Li, J. Ye, S. Ma, and G. Le. 2011. "Ultrasensitive chemiluminescent immunoassay of Salmonella with silver enhancement of nanogold labels." *Luminescence* 26 (2):136-41. doi: 10.1002/bio.1196.

Ware, W. R. 1962. "Oxygen quenching of fluorescence in solution: An experimental study of the diffusion process." *The Journal of Physical Chemistry* 66 (3):455-458. doi: 10.1021/j100809a020.

Wasselin-Trupin, V., G. Baldacchino, S. Bouffard, E. Balanzat, M. Gardès-Albert, Z. Abedinzadeh, D. Jore, S. Deycard, and B. Hickel. 2000. "A New Method for the Measurement of Low Concentrations of OH/O2-

Radical Species in Water by High-LET Pulse Radiolysis. A Time-Resolved Chemiluminescence Study." *The Journal of Physical Chemistry A* 104 (38):8709-8714. doi: 10.1021/jp000462z.

Wendler, J., A. Hoffmann, G. Gross, H. A. Weich, and U. Bilitewski. 2005. "Development of an enzyme-linked immunoreceptor assay (ELIRA) for quantification of the biological activity of recombinant human bone morphogenetic protein-2." *J Biotechnol* 119 (4):425-35. doi: 10.1016/j.jbiotec.2005.04.021.

Wu, Yafeng, Yafeng Zhuang, Songqin Liu, and Lin He. 2008. "Phenylboronic acid immunoaffinity reactor coupled with flow injection chemiluminescence for determination of α-fetoprotein." *Analytica Chimica Acta* 630 (2):186-193. doi: doi.org/10.1016/j.aca.2008.10.010.

Wylie, R. S., and D. H. Macartney. 1992. "Self-assembling metal rotaxane complexes of alpha-cyclodextrin." *Journal of the American Chemical Society* 114 (8):3136-3138. doi: 10.1021/ja00034a069.

Xu, S. L., and H. Cui. 2007. "Luminol chemiluminescence catalysed by colloidal platinum nanoparticles." *Luminescence* 22 (2):77-87. doi: 10.1002/bio.929.

Yang, F. Q., S. P. Li, Y. Chen, S. C. Lao, Y. T. Wang, T. T. Dong, and K. W. Tsim. 2005. "Identification and quantitation of eleven sesquiterpenes in three species of Curcuma rhizomes by pressurized liquid extraction and gas chromatography-mass spectrometry." *J Pharm Biomed Anal* 39 (3-4):552-8. doi: 10.1016/j.jpba.2005.05.001.

Yeshion, T. E. Stanley, P. E. Kricka, K. J. 1991. *Bioluminescence and Chemiluminescence: Current Status*. Vol. 7. New York (Ed) John Wiley.

Yu, S., A. Perálvarez-Marín, C. Minelli, J. Faraudo, A. Roig, and A. Laromaine. 2016. "Albumin-coated SPIONs: an experimental and theoretical evaluation of protein conformation, binding affinity and competition with serum proteins." *Nanoscale* 8 (30):14393-14405. doi: 10.1039/C6NR01732K.

Zana, R. 1986. *Surfactant Solutions: New Methods of Investigation*. New York: Marcel Dekker.

Zhang, H., T. Shibata, T. Krawczyk, T. Kabashima, J. Lu, M. K. Lee, and M. Kai. 2009. "Facile detection of proteins on a solid-phase membrane by direct binding of dextran-based luminol–biotin chemiluminescent polymer." *Talanta* 79 (3):700-705. doi: doi.org/10.1016/j.talanta.2009.04.056.

Zhang, Z. F., H. Cui, C. Z. Lai, and L. J. Liu. 2005. "Gold nanoparticle-catalyzed luminol chemiluminescence and its analytical applications." *Anal Chem* 77 (10):3324-9. doi: 10.1021/ac050036f.

Zhong, D., A. Douhal, and A. H. Zewail. 2000. "Femtosecond studies of protein-ligand hydrophobic binding and dynamics: human serum albumin." *Proc Natl Acad Sci USA* 97 (26):14056-61. doi: 10.1073/pnas.250491297.

CONTENTS OF EARLIER VOLUMES

Advances in Chemistry Research. Volume 53

Advances in Chemistry Research. Volume 52

Advances in Chemistry Research. Volume 51

Advances in Chemistry Research. Volume 50

Advances in Chemistry Research. Volume 49

INDEX

#

A

B

C

D

G

H

M

N

O

P

Q

R

S

T

U

V

W

X

Y

Z

π